SEAWATER CONCENTRATE

For

ABUNDANT AGRICULTURE

By

Arthur Zeigler

Dedication

This book is dedicated to Doctor Maynard Murray, medical doctor and research scientist 1910-1983. During his years of research, from the late 1930's to his passing in 1983, he did wonderful pioneering work proving conclusively that seawater contains bioactive principles important for optimum growth and health of plants and animals, both in the sea and on the land.

INTRODUCTION

Since the beginning of the so-called Green Revolution that followed the Second World War, a new and modern agricultural paradigm has reigned supreme in the minds of the farming community. This new design was based on the premise that nature is flawed, and that we can fix it.

The new paradigm was based on the use of soluble fertilizers to rapidly stimulate plant growth, to produce more bins and bushels of crops on ever greater acreages. This process resulted in weakened plants grown on poor land that could now be induced to produce. These weakened plants needed to be protected, with chemicals, from the natural messengers sent to take them out of production.

The fix involved a battle with the elements of the natural system. The promise of the modern vision was the elimination of fungal and bacterial pathogens, reduction of insect pests, and eradication of weeds from the farmers' fields. As we can see today, with the perfected vision of hindsight, these promises have not been fulfilled. Indeed, today agriculture faces even more severe challenges than it did in the past. Diseases, insects, and weeds have all continued to flourish, proliferate and grow stronger, rapidly developing resistance to an endless succession of newer and ever stronger pesticides.

All of this came at a cost that was greater than many realized. The true cost begins with the purchase price for the bill of goods sold to the farmer. Then there is the seldom considered environmental cost of producing these materials and bringing them to the farm. Even greater and more disastrous, has been the cost of soil degradation. Soluble fertilizers added to soils solubilize the organic substances in soil, and release them into the soil solution. This reaction results in rapid plant growth and greater productivity. In time, however, these soils lose their organic materials and subsequently, ever increasing quantities of fertilizers are needed to produce the same results. As these soils degrade, the minerals contained in the soil matrix rapidly become solubilized and leach out of the soil profile, or they become tied up in a form that is unavailable to plants.

This situation has resulted in crops that contain much lower levels of minerals and vitamins than were found prior to the instigation of these new methods. It has been well established that plants require a broad diversity of trace minerals. These minerals function as enzyme cofactors that are necessary to properly execute even the most basic biological functions. Many plants require a minimum of 59 different mineral elements to have a completely functioning enzyme system for optimal development. These trace elements have become severely depleted in our soils and in our crops as a result of recent agricultural practices.

There is much discussion today concerning agricultural sustainability. I believe we are beyond the point for sustainability. Our soils have become too degraded and depleted. We are too far down hill to be sustainable. We need a regenerative agriculture. A regenerative agricultural paradigm needs to be focused on rebuilding soil and plant health to its formerly high level of inherent potential. When this level is reached we can then begin to have discussions about agricultural sustainability.

Rebuilding agricultural potential is where ocean mineral supplements make an entry. In this book the author describes the benefits of using ocean mineral supplements, an important tool in the toolbox to rebuild agriculture.

In our agricultural consulting business we have used ocean mineral supplements widely, especially SEA-CROP®, and we continue to experience profound results with all types of plants and crops. These results cannot be easily replicated using other substances. Seawater mineral concentrates are unique.

Let us make use of all the opportunities available to us as we move forward in rebuilding our food production resources!

John Kempf -agronomist
Advancing Eco-Agriculture LLC
www.growbetterfood.com

PROLOGUE

The task I set for myself in writing this book was to present the reader with a history of seawater concentrate research as it applies to both plant and animal life.

The period to be covered begins with Dr. Maynard Murray's initial work in the late 1930s and moves forward through his life's work and then encompasses the author's research. It will look at the similarities and differences between the two.

Of necessity, there is quite a lot of statistical data set forth in the book and I have endeavored to present it as clearly as possible. With all of the data from field trials, the book is dense with information.

This is not a scholarly work so in an effort to keep the text from being overly dry, I have not burdened the pages with footnotes but instead have struggled to make the writing as clear as possible. Notes for each chapter have been placed at the back of the book after the index.

If at any point confusion is created in the mind of the reader, please accept in advance the apologies of this author who is a writer by necessity and not by occupation.

One purpose of this book is to review the types of seawater concentrate available for use in agriculture and document the benefits they offer.

Another aim is to identify, for correction, certain destructive agricultural practices currently in use.

To the best of my knowledge there is no other source where all this information is brought together. With this in mind I hope that this book can be of service to the reader and to the public at large through the implementation of seawater concentrate usage in agriculture together with all the blessings that will bring for mankind.

CONTENTS

Chapter One

What is Seawater?

"The cure for anything is salt water--sweat, tears, or the sea."~ Isak Dinesen

Seawater is water from a sea or ocean. Everything soluble, given enough time, eventually ends up in the sea.

The rivers of the world dump billions of tons of minerals into the oceans each year and have for eons. Undersea volcanoes, hot springs and volcanic eruptions on land have also added billions of tons of elements. Ninety percent of all volcanic activity on Earth occurs in the oceans. Additionally seawater is a biological soup of active organic substances.

Life on Earth is totally dependent on water and water comprises 60% - 70% of all living matter. The surface of the Earth is 71% covered by the water of the world's oceans.

According to the National Oceanic and Atmospheric Administration (NOAA), 50% of all the species of life on Earth are found in the sea. In fact, NOAA states that the oceans represent our planet's largest habitat, containing 99% of the planet's living space.

Dr. Maynard Murray, the medical doctor and seawater researcher to whom this book is dedicated, stated that fully 85% of all life that exists on Planet Earth exists in the sea.

Those who do not live at the seashore tend to have limited understanding of the vastness of the oceans and how they contribute to life on land.

As a consequence of covering nearly three quarters of Earth's surface, the oceans are the planet's largest interceptors of solar wind and transducers of solar energy. Transduction is the transfer of energy from one form to another. Solar wind is comprised of substances and energies that are poorly understood and so it follows that what occurs when those substances and energies are intercepted by the oceans is also not well understood.

One method of solar energy transduction in the oceans is the creation of organic matter through photosynthesis by phytoplankton and photosynthesizing marine bacteria. This may be described as the transmutation of radiant energy into matter. Marine phytoplanktons are so abundant they are credited with the manufacture of nearly 50% of the oxygen found in Earth's atmosphere.[1] Each of these individual plankton are so tiny that without a microscope you would never know that they exist.

Certain marine bacteria prevalent in ocean water, at the average rate of one billion per liter, are responsible for much of the rest of atmospheric oxygen produced by photosynthesis. Many of these bacteria are born unable to photosynthesize until they become invaded by marine viruses that instruct them to manufacture a pigment that performs a function similar to chlorophyll. The pigment, named proteorhodopsin, is capable of intercepting photons, so photosynthesis can take place. Marine viruses are prevalent in seawater at the rate of 10 billion per liter.[2] Their contribution is so important that we should give thanks to them for the oxygen in every breath we take.

The existence of microscopic marine organisms allows mammalian life on land to exist. And, as a byproduct of their life cycle, they also create in the oceans, a complex "stew" of organic compounds such as RNA, DNA, proteins and polysaccharides.

In addition to the compounds that are byproducts of photosynthesis, the waters of the oceans contain 89 elements in measurable amounts. Every element that occurs in nature is found in the sea. There are also many more minerals in compound form, such as sodium chloride, that have been leached out from the vast area of land covered by oceans.

An enormous amount of soluble organic materials are contributed by all the waters that flow from the land to the sea as well. If the reader has ever seen an otherwise clear stream that had a yellowish or brown coloration, they have seen fulvic acid on its way to the sea. Fulvic acid contains over 50,000 different organic substances. Fulvic acid has been detected in all seawater.

In total, the seas of the world are estimated to contain as much as fifty quadrillion tons (50,000,000,000,000,000) of dissolved solids. If these dissolved minerals could be removed from the seas and spread on the land as dried solids they would cover all the land surfaces of planet Earth with a layer of minerals five hundred feet thick.[3]

Seawater on the average has a salinity of about 3.5% or 35 grams to the liter of mixed chemical salts.

10 Most Common Elements in Seawater			
Oxygen	85.84%	Sulfur	0.091%
Hydrogen	10.82%	Calcium	0.04%
Chloride	1.94%	Potassium	0.04%
Sodium	1.08%	Bromine	0.0067%
Magnesium	0.1292%	Carbon	0.0028%

The principle dissolved solids in seawater are sodium chloride (common salt), the calcium salts, calcium carbonate and calcium chloride, potassium sulfate and the magnesium salts magnesium chloride, magnesium sulfate and magnesium bromide.

There are 80 other elements present in seawater in trace and ultra-trace amounts.

In addition to all of the elements and complex mineral salts, there are phytoplankton, zooplankton, microbes and the byproducts of their lifecycles.

We are completely unaware of photosynthesizing marine bacteria when we go swimming in the ocean but they may be the largest single biomass on the planet, far exceeding the forests of the world.

Dr. Maynard Murray stated, "A cubic foot of ocean water sustains many more times the number of living organisms, plants and animals than does the equivalent amount of soil. Seawater is literally alive..."

The mineral content of seawater is fairly uniform throughout the oceans of the world. The exception to this is the element Phosphorus that forms, together with calcium and oxygen, the complex mineral called Apatite. Being insoluble, Apatite precipitates out of solution, sinking to the bottom of the sea. The Pacific Ocean, due to the volcanic activity of the "Ring of Fire" contains twice the Phosphorus content as the Atlantic Ocean.

When deep-sea ocean currents upwell against the continental shelves of the land masses of the world, phosphorus is brought up from the depths and re-enters the food chain. As its absence is a limiting factor, the presence of phosphorus enables a burgeoning of the populations of the photosynthesizing plankton known as **phytoplankton**. The name comes from the Greek words *phyton* meaning plant and *planktos* meaning wanderer or drifter and this name very well describes how these microalgae spend their lives.

Dwelling in the sea, as they do, plankton spend their lives in the presence of large amounts of sodium, as in sodium chloride. To prevent too much sodium from entering their cells and inhibiting their function, they coat themselves with a mucopolysaccharide exudate named dimethylsulfoniopropionate, or DMSP. This has an osmoregulatory function that protects the cells from changes in salinity and temperature.

While this is obviously important to the phytoplankton, why is this important to us? It turns out that DMSP is produced by phytoplankton in such massive amounts that it has a major effect on planetary weather as well as being a major player in both the carbon cycle and sulfur cycle in the seas.[4]

DMSP sloughs off from phytoplankton during their lifetimes and when they are grazed on by zooplankton. The name **zooplankton** comes from the Greek word *zoon* meaning animal and *plankto*s meaning a wanderer or drifter. These can be tiny microorganisms or creatures as large as jellyfish and they include, in addition to bacteria, the larval forms of sea urchins, starfish, shrimp, shellfish and most fish.

In other words phytoplankton are plants that earn their living by intercepting photons from sunlight and transforming that energy into carbohydrates and proteins. They are at the bottom of the marine food chain. They are grazed on by zooplankton, just as the herbivores on land graze on grass and other plants.

As a consequence, DMSP is released into the seawater and may be prevalent in such abundance that it can spontaneously polymerize and turn square miles of seawater into a gelatinous mass. This happened recently in the Arctic Ocean to the great consternation of the mystified news media although there were probably many marine biologists who knew what had happened as soon as they heard of it.

Most of the DMSP released into seawater as a byproduct of the phytoplankton lifecycle is immediately fed upon by bacteria and retained in the marine food chain. There it provides nearly all of the sulfur required in the diet of these microbes. A much smaller portion of the DMSP is acted upon by an enzyme called DMSP-lyase produced by other bacteria.

One of the products of this enzyme action is the volatile compound dimethylsulfide (DMS). Some of the DMS thus produced is oxidized to DMSO, a solvent compound that many, including this author, have used as a liniment. Although marine microbes metabolize much of the DMS, each year about 50 million tons of this chemical compound escape, as a gas, to the atmosphere where it acts as a cloud seeding agent. It is this cloud seeding activity of DMS, derived from DMSP, which makes the humble phytoplankton a major player in planetary weather.

I live in the coastal rainforest of Washington State where we enjoy 85 inches of rainfall in the average year. This coastal rainfall is made possible by the westerly prevailing winds that bring the DMS seeded cloud formations on shore to release their precipitation. This same phenomenon is responsible for the deserts that exist further inland where, after the costal rainfall, no water vapor is left in the atmosphere as it is carried eastward away from the marine environment.

For all that seawater is a major habitat for microbes; we are only beginning to learn about marine microbes. The only bacteria that can be studied are the ones that can be raised in laboratories and marine bacteria are notoriously difficult to culture. Only a few strains have been successfully cultured and extensively studied.

Our ignorance concerning bacterial life in the oceans is as vast as the oceans themselves.

The marine environment sustains a wide diversity of plants and animals. Close to 200,000 species of marine algae, animal bacteria, fungi and viruses have been identified with perhaps four times this number of organisms known to be present but as yet unidentified.

In addition to the great array of life in the oceans, there are many organic chemicals present, derived from the life cycles of plants and animals both on the land and in the sea. For example all ocean water, without exception, contains fulvic acid. Fulvic acid is estimated to be composed of over 50,000 different organic compounds. Due to its complexity, no one knows all of its components and that will probably remain true far into the future. Fulvic acid contains fragments of long chain organic compounds such as DNA, RNA, vitamins, hormones and enzymes that have degraded to the point that they become stable and then persist in nature for long periods of time.

Taken together, both the extensive mineral content and the many organic compounds present make seawater a stew of considerable complexity. This is not just salt water. The fluid that covers nearly three fourths of our planet is home to the major part of life on Earth and it is the activity of life in the seas that makes life on land possible.

Of these mineral and carbon components, sodium chloride (sea salt) is vital to human life and has been obtained from seawater since the earliest prehistory of the human race. One can argue that this is the first known use of concentrating seawater for human use. Since reviewing seawater concentrate use in agriculture, for both plants and animals, is the primary purpose of this book, let's review some of the methods used to concentrate seawater.

Seawater can be concentrated and there are a number of ways to accomplish this.

Sodium chloride can be obtained from seawater by simple evaporation. It is an interesting fact that as seawater is evaporated and becomes a supersaturated solution, sodium chloride crystallizes out of solution while the other minerals remain in solution. This provides a method for extracting sodium chloride and separating it in relatively pure form as sea salt.

Because the concentrated minerals that it contains have a very bitter taste, the brine remaining after the sea salt has been extracted is referred to as bitterns. This is one way of producing a seawater concentrate that excludes most of the sodium chloride but unfortunately bitterns are not particularly bioactive when used in agriculture. A substance is considered to be bioactive when it has an effect on living matter. It is possible that this lack of bioactivity is because the organic component becomes degraded due to ultraviolet light exposure during the lengthy solar evaporation process.

A more modern method of producing a seawater concentrate is reverse osmosis. This method is used to produce drinking water and also produces, as a byproduct, a mineral concentrate that contains all of the organic and inorganic components. This method will produce a bioactive concentrate but the mineral content is mostly sodium chloride, which limits its efficacy and utility.

Salt deposits from geological eras in the past exist in many parts of the world and such deposits have been mined for thousands of years. The Hallstatt salt mine in Austria has been a source of sodium chloride for 7,000 years. Except for a few species of plants that are sodium chloride lovers, this type of salt has little agricultural use except as a mineral supplement for animals.

In August of 2007 the United States Patent Office issued, in the author's name, patent number 7,261,912 entitled **Method of producing useful products from seawater and similar microflora containing brines**.

Among other claims the patent claimed the following:

"A method under claim 1 for producing a new composition of matter from seawater or brines containing aquatic microflora comprising: the precipitates produced as in claim twenty six comprise a new composition of matter, an intimate mixture of the naturally occurring bivalent minerals of the brine co-precipitated and condensed together with, incidental trace minerals, substantially all naturally occurring carbon-based chemicals and carbon-based particulate matter along with entrained water, water of hydration, phytoplankton, phytobacter, nanobacter,

viruses and virus-like particles together with the byproducts of their life-cycles, said new composition of matter containing up to one billion bacteria and ten billion viruses and virus-like particles per tablespoon-full, the entrained waters, in the form of clathrates, quasi-clathrates and hydrogels, being an inextricable constituent of the new composition of matter."

The method used in this patent involves separating out and concentrating the trace and ultra-trace elements together with the organic matter. The sodium chloride is left behind in solution so the other minerals are safe to use on soil without concern about a salt buildup.

This seawater mineral concentrate is bioactive and effective as a soil amendment and foliar treatment for use in agriculture. My company, Ambrosia Technology LLC, manufactures a unique and potent seawater concentrate that we market under the trade name SEA-CROP ®.

In this chapter, we have covered a few of the facts that show seawater is an unusually complex and unique resource, that it contains bioactive principles and that they can be concentrated.

We have seen that there are various methods of producing seawater concentrates that result in products with widely different characteristics and bioactivity in regards to agricultural use. We shall explore this in depth further on in the book.

The next chapter recounts the history of more than a century of research by various investigators who examined and documented the bioactivity of seawater concentrates as applied to plant and animal life.

Chapter Two

A History of Seawater Concentrate Research

"My research clearly indicates the reason Americans generally lack a complete physiological chemistry is that the balanced, essential elements of the soil have eroded to the sea; consequently, crops are nutritionally poor, and the animals eating these plants are, therefore, nutritionally poor . . . We must alter the way we grow our food, the way we protect our plants from pests and disease, and the way we process our food." ~ Dr. Maynard Murray

Scientific researchers have been examining the effects of seawater, seawater extracts and seawater concentrates on terrestrial life since the late 1800s. In that era, Dr. René Quinton, the renowned French doctor, biologist, biochemist and physiologist observed that seawater had remarkable similarities to blood plasma.

In 1904, Dr. Quinton, who has been called the French Darwin, published the book, Sea Water, organic medium, in which he described his research into the relationship between blood plasma and seawater. He did experiments with dogs by removing most of their blood and replacing it with seawater extract. Not only did the dogs survive, they remained in an excellent state of health. Dr. Quinton went on to develop an injectable form of purified seawater that was used as a replacement for blood plasma during World War I thus saving thousands of lives with seawater.

This brilliant researcher believed strongly that seawater should be treated as an organic substance, not just a collection of inorganic chemicals in solution.

Because of his unique research and lifesaving medical discoveries, when Dr. Quinton died in 1925, as a result of wounds received during WWI, he was a French national hero. People flocked to his funeral from all over the world in recognition of his work and humanitarian contributions.

As early as the late 1930s, Dr. Maynard Murray noticed profound effects when seawater was applied to living organisms.

Dr. Murray became fascinated with seawater in the early 1930s as he became aware that there are no chronic diseases among fish and animal life in the sea that compare to the diseases to which terrestrial animals, including mankind, are subject to. Furthermore, he notice that fish and marine animals have relative life spans that are much greater and they are not subject to ageing processes to the same degree as terrestrial life. He set out to find what it was in seawater that seemed to make the waters of the seas a veritable fountain of youth. As a result, throughout his entire adult life, he engaged in research with seawater and seawater concentrate.

He theorized the apparent difference in disease resistance and vitality between life on land and in the sea is due to mineral deficiencies in our soil and food. He visualized an endless cycle wherein continents rise from the sea rich with minerals. The constant effects of climate: freezing, thawing, rainfall and erosion combined with mankind's historically poor stewardship of the land and increasing acidic rain cause topsoil minerals to go into solution. These mineral solutions enter streams and rivers that subsequently flow into the sea. Dr. Murray concluded that these minerals hold the key to human health and that it made perfect sense to recapture them and restore them to our soils.

Initially, his experiments used diluted seawater on soils and crops. In an effort to reduce shipping expense he decided to try working with seawater evaporated to complete dryness. Then only the 3.5% of seawater that is the mineral content would need to be shipped. He called these dried minerals sea solids and used them, during many years of extensive and well documented research, on all types of crops and soils.

The results were consistently the same: the plants flourished, matured more rapidly, were healthier, more disease and drought resistant and produced outstanding taste along with greater yields. In assays testing for nutrients, foods grown with Dr. Murray's sea solids had significantly more minerals (ash content), vitamins (25% more vitamin C in tomatoes, 40% more vitamin A in carrots) and sugars. In addition, he witnessed the same amazing results in all types of livestock and poultry that were offered feed grown in soil enriched with his sea solids. Physiologically, animals were healthier, gained weight more rapidly and reached maturity sooner. The main difficulty with Dr. Murray's system was that it required application of 1,500 pounds per acre of dried sea salt to achieve the desired results.[1]

As with so many pioneers, he was ahead of his time and had available to him many fewer pieces of the puzzle than we do now. Researchers in the last few decades have accumulated much knowledge concerning the teeming microfloral life in fertile soils. During his lifetime the degree to which plants and soil microflora live in symbiosis and their constant exchange of chemical compounds was unknown. That lack of information may have led to his rigid insistence that plants could only benefit from inorganic chemicals and not from organic inputs.

As a result, he completely ignored the very important organic bioactive components of seawater and seawater concentrates. These organic compounds working in concert with the 89 elements contained in seawater have a profound effect on both soil microflora and plants.

Dr. Murray felt so strongly about this that in the first paragraph of his June 1, 1963 patent application he specifically excludes the organic component of seawater from the patent.[2]

About 12 years ago, I took up the challenge to extract from seawater the bioactive principles without the salt. Proceeding from the work of Dr. Murray and guided by a test for bioactivity recommended by Dr. Philip Callahan, I entered onto a path of research that was to culminate with the creation of the product SEA-CROP®. The first goal was to extract the bioactive principles while leaving behind most of the water and sodium chloride. It seemed that this would be the only way a seawater extract would become economical and have a broad-based agricultural application.

Eventually the research resulted in a bioactive seawater concentrate with reduced salt and loaded with trace minerals. It performed well, compared to Dr. Murray's system, but instead of 1,500 lbs per acre only a few gallons per acre of seawater concentrate were needed.

Early experimentation with this concentrate showed that if the extract was allowed to dry out completely much of the bioactivity was lost. Further testing revealed that if the extract was dried in an oxidizing atmosphere, at a temperature sufficient to burn off all contained carbon, the resulting pure mineral concentrate had little observable bioactivity at the same application rate previously used.

As a result of these experiments, a working hypothesis was developed. It was postulated that active organic substances in seawater, working together with the plentiful trace minerals, were responsible for the observed bioactivity when the extract was used as a plant and soil stimulant.

As mentioned previously, Dr. Murray observed that a cubic foot of seawater contains considerably more living organisms than an equivalent amount of soil.

Total Organic Carbon (TOC) assays of both our extract and the stripped seawater showed that we were indeed concentrating substantially all of the organic material from seawater.

As no record of a claim to prior discovery could be found for this method of removing the trace minerals from seawater, leaving the salt behind and concentrating all of seawater's organic content, a patent for the process was applied for and US Patent # 7,261,912 was eventually granted on August 28, 2007.

Seawater concentrate produced by the patented method did meet the original goal of extracting from seawater most of the bioactive principles while leaving behind most of the water and sodium chloride.

Further research and several improvements in the process led to discoveries that resulted in SEA-CROP®, the seawater concentrate Ambrosia Technology has been making for the last six years.

SEA-CROP® is certified by the Washington State Department of Agriculture (WSDA) as fully meeting the requirements of the National Organic Program. It has been authorized by the WSDA, the California Department of Food and Agriculture and Oregon Tilth for unlimited use in production of organic food crops.

What Does Seawater Concentrate Do?

Dr. Maynard Murray found that using seawater concentrate enhanced both yield and quality for every crop on which it was used. In addition, when those harvested crops were feed to chickens, pigs and cattle, they reached maturity much sooner than control animals fed conventionally raised crops. They were also much healthier and disease resistant than the controls.

SEA-CROP® has shown the ability to give the same kinds or results as those that Dr. Murray achieved but instead of needing 1,500 pounds per acre, only a few gallons per acre were required to achieve the same results.

Cattle on pasture that has been treated with SEA-CROP®, or fed hay made from a SEA-CROP® treated field, responded the same way as the cattle in Dr. Murrays's trials. Due to the fact that SEA-CROP® is 95% sodium chloride reduced, it can also be beneficially given directly to animals in their water or on their feed. The animals can be given enough trace minerals and other bioactive factors to have a beneficial effect without the stress of ingesting excess sodium chloride.

While it is possible to measure changes in the physiology of plants and animals that have been treated with seawater concentrate and compare the measurements to those of untreated controls, this does not tell us why the changes occur. Such comparisons only quantify the changes so that we can put numbers on them. The exact mechanisms by which seawater concentrates work their magic is unknown.

We can however make some qualifiable and working assumptions as to what is happening.

1. Providing a buffet banquet of 89 elements for mineral cofactors enables enhanced enzyme formation by plants, soil bacteria, fungi and soil fauna of all kinds from the most microscopic all the way up to the macroflora, earthworms, birds and small mammals that feed on them. Recall that enzymes are used by bacteria that comprise the digestive systems of both plants and animals whether they are internal, such those in our gut, or external like those of plants, fungi and bacteria.

2. Increased efficiency of a plant's solar antenna due to increased carotenoid production. This could lead to a greater ability to photosynthesize and transfer solar energy to the plant for beneficial use such as production of ATP, other metabolites and simple sugars exudates from the root hairs to the symbiotic entities in the rhizosphere.

3. Increased growth hormones such as auxin and messenger molecules such as cytokinins. Plants that have been treated with SEA-CROP® often develop root systems that are 20% to 30% larger than untreated controls. Auxin is a growth hormone that is manufactured in the solar factories in the top part of a plant. It is responsible for new root formation and must travel down through the plant in order to reach the roots and be effective there. In the normal course of events, auxin is consumed in other biological operations throughout the plant so that only a certain amount is left by the time that it finally arrives at the root tips where it is to initiate new growth. If additional auxin is produced by the plant, due to increased radiant energy adsorption, then it may be possible that more is unconsumed and available when it completes its journey from the top of the plant to the outermost root tip in the rhizosphere.

4. Increased metabolite production of all kinds in the top part of the plant would mean increased root exudates in the rhizosphere which would mean more food

for an increased population of microflora helpers in the plant's rhizosphere garden.

These are working assumptions that may change as we acquire more knowledge. However, there is reasonable certainty that some if not all of them will be confirmed as further research takes place. Due to the great complexity of the symbiotic biological systems involved and the complexity of the seawater concentrate itself, it may be a long time before full understanding is achieved. In the meantime, we can give thanks that such a wonderful aid for agriculture is available for us to use. At this point, we really do not need to know how it works; only that it *does* work.

Chapter Three

The Benefits of Seawater Concentrate

"Go to the ruins of an ancient and rich civilization in Asia Minor, North Africa or elsewhere. Look at the unpopulated valleys, at the dead and buried cities, and you can decipher there the promise and the prophecy that the law of soil exhaustion holds for all of us. Depleted ....by constant cropping, land could no longer reward labor and support life." ~ V.K. Simkovitch

There are many reasons to use seawater concentrate for agriculture. Among these reasons are increased yield at harvest, improved soil tilth, increased soil microflora, better drought tolerance, increased nutrient density (such as vitamin and mineral content which produces better flavor, keeping qualities and health for the consumer), improved plant health and overall vigor that makes plants more disease and insect resistant.

While many of these benefits are of primary importance to the gardener, the farmer must first of all look to the bottom line and that means that his primary interest must be in yield at harvest. Each planting season he rolls the dice and prays that the promise held in the seed that he plants in the ground can withstand the extremes of drought, flood, wind, and hail, chill and heat so that come harvest time there is sufficient yield to sustain him until the following planting season.

Next for the farmer, after yield per acre, comes the health of the soil. While yield is the short term yearly focus, in the longer term, if the soil itself is degraded or destroyed, yields cannot be improved or even long maintained.

For the fruit and vegetable grower, keeping quality is of considerable importance and improved flavor is certainly a bonus.

For the farmer or rancher who raises forage and feed for use in feeding animals on his own land, nutrient density also becomes an important consideration because animals fed on low nutrient food must struggle to grow properly and maintain good health.

For a farmer selling his crops off the farm, nutrient density may have no importance. It is another matter entirely for the end user of the crop. Recently I spoke with a gentleman who operates a feed lot in Arkansas. He stated that with the feed grains now available to him, especially the GMO crops, it has become difficult to get cattle to gain weight properly. The feed contains calories but calories alone are not enough.

For the moment let's set aside all of the other benefits that experience has shown seawater concentrates provide and look only at the one benefit that has historically been of primary importance to the farmer, yield at harvest.

Dr. Maynard Murray, medical doctor and research scientist, pioneered the use of seawater and seawater concentrate for agriculture. Starting in the late 1930s and continuing until his passing in 1983 he performed well documented trials with plants and animals. We can review his findings in detail. Although, in the light of recent discoveries, some of his theories are dated, his work, taken together with current research, shows an impressive history of documented crop yield increases over the greater part of a century.

Dr. Murray together with the investigative journalist Tom Valentine produced a book in 1976 entitled Sea Energy Agriculture. This book published by Acres U.S.A., is still in print and is available from the publisher's website at www.acresusa.com. It is highly recommended for anyone who wishes to review the origin of the sea mineral for agriculture movement.

In his book, Dr. Murray relates that he started his seawater experiments in 1938. With the help of the U.S. Navy he was able to obtain seawater from all the oceans of the world in carload lots that were delivered to him at the University of Cincinnati Medical School, where he was directing research into the effects of seawater on plant and animal life.

Over the years, he conducted agricultural field trials in the states of Florida, Illinois, Massachusetts, Ohio, Pennsylvania, South Dakota and Wisconsin.

By the mid 1950's, Dr. Murray was no longer using raw seawater due to the difficulty and economics involved in shipping and application of the bulky unconcentrated material.

He had by this time located a deposit of impure sea salt on sun baked salt flats located on the West side of the Gulf of California also known as the Sea of Cortez. About 125 miles South of El Centro, California and the Mexican border town of Mexicali lies the small resort town of San Felipe. It has beautiful sandy beaches where one can go swimming in the sun warmed waters of the Sea of Cortez.

About 15 miles north of town, there is a solar salt works where sea salt is made. Starting there, and extending much of the way north to Mexicali, are salt flats in an area that has in the past been repeatedly inundated by seawater that then partially evaporated and left behind a portion of its mineral content. Although this material is mostly sea salt it does have some trace mineral content and also some bioactivity when applied in sufficient quantity. Dr. Murray named this salt material *Sea Solids*.

This is the material that Dr. Murray was using when he conducted agricultural testing during the mid 1950s. The following information and tables showing harvest yield results are typical of field trials that he performed at that time.

In 1954, at the Ray Heine and Sons Farm 11 miles west of Elgin, Illinois an application of 2,200 pounds per acre of Dr. Murray's sea solids gave an18.4% increased yield of oats against the yield of oats produced in an adjacent portion of the field that was not treated with the sea solids. As part of this same test, a 40 acre field was fertilized normally and then 30 acres of that field were treated with sea solids at the rate of 2,200 pounds to the acre. Then, the entire field was planted to corn. At harvest the treated portion of the field gave a yield 13.3% greater than the untreated portion.[1]

In another test on corn in the year 1970 in southern Wisconsin, a 40 acre field was fertilized normally and then a portion of the field was treated with sea solids at rates varying between 100 pounds to 1,500 pounds per acre before the entire field was planted to corn.

At harvest, it was noted that there was a yield increase in all of the sea solid treated portions of the field and there was a direct relationship between the amount of increased yield and the amount of sea solids applied. The portion of the field that received 1,500 pounds to the acre of sea solids gave an increased yield of 33.9% more bushels to the acre than the portion of the field that only received normal fertilization.

It is interesting to note that the sea solid corn had a greater test weight, giving 57.5 pounds to the bushel compared to 53.5 pounds to the bushel for the untreated control. So in addition to 33.9% more bushels to the acre, the greater weight per bushel adds another 4% to the increased yield for a total of 37.9% increase.

Furthermore, the untreated control had a moisture content of 25% compared to a moisture content of 20% for the "sea solid" treated corn. The increased yield having 20% less moisture content compared to the control means that the actual increased harvest of nutritional matter was even greater than 33.9%. If the corn harvested from the untreated portion of the field was dried so that it had the same 20% moisture content as the corn from the treated portion of the field, an additional 6.7% difference would tally for the treated harvest giving a grand total of 44.6% greater harvest by weight.

This increased dry matter content, when crops are treated with seawater concentrate, is a phenomenon that this author and others have witnessed many times over the years when trialing SEA-CROP® on a variety of plant species. The increased dry matter weight comes from additional minerals taken up by the plant as well as a greater content of metabolites such as additional carbohydrates, vitamins, hormones and proteins.

It was not only corn that showed an increase in dry matter content. Dr. Murray tested other crops during the mid 1950s. All six crops listed below showed increased dry matter content as represented by ash. **Ash** consists of the weight of the elements after all organic material has been burnt out of the sample.

Samples	**Ash Weight in % Solids**		**% of increase**
	Control	Treated	
Onions (Bulb)	13.6	14.2	4.4%
Oats	87.7	87.8	0.1%
Sweet Potatoes	28.8	31.2	8.3%
Tomatoes	4.8	5.7	18.7%
Soy Beans	73.9	84.7	14.6%

Sea Energy Agriculture

The reader has no doubt noticed the considerable amounts of sea solids that Dr. Murray used in conducting these tests. He applied the sea solids in amounts ranging from 100 pounds per acre all the way up to 3,000 pounds per acre. He stated in his book that these amounts were well tolerated and that when used at the 1,500 pound per acre rate, no further application was necessary for several years.

Using these sea solids was no doubt a significant improvement over shipping bulky seawater inland. However, because of the large sodium chloride content it was necessary to use large amounts of sea solids in order to obtain enough marine trace minerals to get the desired seawater effect. While this worked in the Midwest and under hydroponic conditions in Florida, one has to wonder if application of such large amounts of sodium chloride would poison the earth in drier areas such as southern California or Arizona. Those areas already have a salt buildup problem to the extent that even using manure can be a hazard because of salt content. Each ton of manure contains 50 to 100 pounds of salt.[2]

Fortunately, SEA-CROP® is a solution to this problem as well as to the shipping and application difficulties. In SEA-CROP® 95% of the sodium chloride is excluded and all of the other minerals are concentrated between 20 to 30 fold, depending on which mineral, as they do not all concentrate at the same rate.

Another advantage that SEA-CROP® has over the sea solids is that, because it is a concentrate of only the active principles contained in seawater, much less of it needs to be used. Application rates are typically only 2 to 4 gallons per acre with a few specialty crops requiring up to 6 gallons to the acre each year.

Even with these low applications rates, SEA-CROP® is able to perform much the same as Dr. Murray's sea solids did in his field trials or better.

The following harvest yield results were obtained during field trials sponsored by third parties. Ambrosia Technology LLC, had no involvement in any of them other than to provide the SEA-CROP® product.

In 2007, a field trial sponsored by Cal-Agri Products was conducted by Dr. Edward McGawley on corn at the University of Louisiana at Baton Rouge with the following results. Two test plots were treated at planting with a single application of SEA-CROP®. One was treated with 2 gallons per acre and the other with 4 gallons per acre. Compared to an untreated control test plot the treated plots yielded increased corn harvests of 58.15% and 94.49%.

Cal-Agri Products had Dr. McGawley conduct additional field trials at Louisiana State University at Baton Rouge in 2008. In this trial, SEA-CROP® was used as a root dip on plug transplants at the dilution rate of one part in fifty (2%). The treatment gave the following increases in marketable fruit:

% Increases Over Control

Plant	% Increase
Peppers	71%
Eggplants	66%
Tomatoes	30%
Strawberries	25%

See graph on page 71

All of the tests at LSU took place in nematode infested soil that was inoculated with nematodes prior to or at planting.

Cali-Agri also sponsored testing by Dr. Abdelhaq Hanafi at the University of Morocco at Agidar.

In 2007, eggplant sets were planted in 5 liter pots and 8 replicates were then treated with a foliar spray of SEA-CROP® that was diluted to a 1% concentration. There was no increase in the amount of fruit but the fruit of the treated plants were 79.5% larger by weight giving a 79% increase in yield over the control.

In 2008, Dr. Abdelhaq Hanafi also conducted tests for Cal-Agri Products on the following crops:

- **Runner Beans:**
  SEA-CROP® was used as a 1% soil drench in the two leaf stage after emergence and as a 1% foliar spray three and six weeks after the soil drench. Total SEA-CROP® usage was three gallons per acre and the increase in marketable fruit was 86%. This test demonstrated that SEA-CROP® improved plant growth, leaf size, root volume and root weight while accelerating early production and improving overall yield.
- **Zucchini:**
  SEA-CROP® was evaluated on 120 plants grown hydroponically in coco-peat using 5 different combinations of soil drench and foliar applications as

compared with an untreated control. The best performance was achieved when SEA-CROP® was applied twice as a 1% foliar application, while the second best was achieved when SEA-CROP® was applied twice as a 0.30% drench. One foliar application of SEA-CROP® at 1% achieved better results than one receiving a soil drench at 0.30% concentration. It was found that SEA-CROP® improves leaf size, number of fruits and total yield, while causing an early flowering and an early production compared to the control. This is illustrated in the graph on page 72 of this book.

In Northern Europe, Ambrosia Technology LLC, products are distributed by Hak Agro Feed. Starting in 2005, this company trialed SEA-CROP® on potatoes for three consecutive years with excellent results. Barend Hak stated, "We had an increase of 28% three years in a row on a one hectare test plot vis-à-vis the control. The potato grown was Bintje. Twice, 3.75 liter per hectare was treated. First spraying was done when the plants were sprouted and fully above the field. Second spraying was at full blossoming."

In 2010-2011, our international distributor, Collé Agriculture, sponsored testing in Malawi, Africa by four different entities at six widely separate sites. The object of the trials was to evaluate the efficacy of SEA-CROP® on a variety of locally important crops including maize, peanuts, soybeans and tobacco. The results follow.

- **Tobacco**
  Two tests performed on tobacco by a research institute gave an average yield increase of 38.5% when SEA-CROP® was applied at the rate 4 gallons per acre split into three applications, one soil drench and two foliar.
- **Peanuts**
  The increased yield obtained by a farmer on a private estate was 42.8% greater than the untreated control. SEA-CROP® was used in this trial at the rate of 4 gallons per acre. This same farmer also achieved increased yields in soybeans 9.3% and maize (corn) 27%.

  Similar results were seen in trials at Bunda Agricultural College and government agricultural research stations of the Ministry of Agriculture & Food Security of the Government of Malawi.

County Line Farm 2012 Winter Trials by Collé Agriculture

In the early part of 2012, Collé Agriculture performed testing at a commercial organic produce farm in Southern California. The testing documented the performance of SEA-CROP® under organic growing conditions.

One aim was to determine if yields increased when the only variable was the addition of SEA-CROP®. Another goal was to determine if produce grown under commercial conditions and treated with SEA-CROP® contained a greater nutritional value.

The crops chosen for testing were baby beets, baby carrots, cauliflower, fennel, spinach and baby Swiss chard. These crops were grown in fields fertilized with hydrolyzed fish powder, chicken manure, and with both macro and micro mineral soil amendments.

A portion of each crop received treatment with SEA-CROP® in addition to the normal fertilization. One part of the treated plants received 2 applications of SEA-CROP®: 1 soil drench and 1 foliar application, totaling 3 gallons to the acre. The other part received 3 applications: one soil drench and 2 foliar applications, totaling 4 gallons per acre. The larger, untreated, portion of each crop acted as the control.

As can be seen in the table below, yields were increased significantly.

Yield % Change from Control

Crop Planted	2 applications % increase	3 applications % increase
Beets	+60.92	+90.13
Carrots	+25.37	+41.57
Cauliflower	+72.9	+89.5
Fennel	+11.26	+38.74
Spinach	+43.0	+70.0
Swiss Chard	+67.31	+45.43
Average for the 6 species	**+46.79**	**+62.56**

For cauliflower and spinach, total yields were used. For the rest, marketable yields, after culling for substandard produce, were used.

The next table of results from the same trial shows comparative dry matter content. Dry matter represents everything in the sample other than water. Dry matter includes protein, fiber, fat, minerals, carbohydrates, etc.

Dry Matter % Change **from** Control (average of 2 methods)

Crop – Dry Matter	**2 applications % increase**	**3 applications % increase**
Beets	+16.46	+5.79
Carrots	-4.56	-0.95
Cauliflower	+13.98	+12.02
Fennel	+0.46	+2.82
Spinach	+4.92	+1.42
Swiss Chard	+2.55	+55.21
Average for the 6 species	**+5.64**	**+12.72**

Dry matter increases indicate enhanced flavor, better keeping quality and greater nutrient density.

Degrees Brix is a measurement of the sugar content of an aqueous solution. During the trial Brix readings of the crops were taken with an optical device called a Brix refractometer. In addition to sugar content, the optical Brix readings include everything that is soluble in the liquid portion of the plant including vitamins, enzymes and hormones.

The following table shows that the treated crops had elevated Brix levels over the untreated controls. These results are a good confirmation of the dry matter readings in the table above.

Brix %Change from Control

Crop Planted	**2 applications % increase**	**3 applications % increase**
Beets	+37.5	+75.0
Carrots	+31.4	+27.9
Cauliflower	+5.5	+11.1
Fennel	+28.6	+57.1
Spinach	+28.6	+23.8
Swiss Chard	-4.7	+15.8
Average for the 6 species	**+21.15**	**+35.12**

As with dry matter content, Brix increases indicate enhanced flavor, better keeping quality and greater nutrient density.

Insect Damage Reduction

Insect Damage Reduction % Change from Control

Crop Planted	2 applications % Reduction	3 applications % Reduction
Beets	0.0	100.0
Carrots	33.3	33.3
Cauliflower	15.0	45.0
Fennel	-	-
Spinach	-	-
Swiss Chard	33.3	33.3
Average for the 4 species	**20.4**	**52.9**

There was no measurable insect damage in the Fennel or Spinach crops.

The results clearly show that in each category: yield, dry matter, Brix, and insect damage reduction, there is a cumulative benefit with additional applications.

The following table shows that protein content increases when plants are treated with SEA-CROP®.

Protein % Change from Control by Nitrogen Analysis

Crop Planted	2 applications % increase	3 applications % increase
Beets	-7.52	+31.58
Carrots	+15.48	+7.74
Cauliflower (mature)	+18.49	+ 21.46
Fennel	+20.91	+47.91
Spinach	+6.48	+8.64
Swiss Chard	+1.47	+1.05
Average for the 6 species	**+9.22**	**+19.73**

In a world hungry for protein these are very significant increases.

All of these third party testing results show that SEA-CROP® is able to perform at least as well as Dr. Murray's sea solids did, or better. This illustrates that the bioactive principles contained in seawater can be greatly concentrated for more economical shipment and ease of application.

In this chapter we have emphasized the economic benefits seawater concentrate may produce for the farmer through increased yields at harvest. Before we go on to the next chapter there is another economic benefit that should be discussed.

In many cases, nitrogen and other inputs may be reduced while maintaining the same or even increased levels of crop production. Depending on the soil and the crop it may be possible to reduce chemical fertilizer inputs, such as nitrogen, thus providing significant savings and improvement in soil microflora populations. Over a 2 year period using SEA-CROP®, one farmer reduced his nitrogen input by 30% while increasing yields above the state average significantly. A case study of a popcorn farmer in Indiana whose field is managed by John Kempf of Advancing Eco-Agriculture illustrates this:

Year	Nitrogen per acre	% reduction	State average yield	This farm's yield	% better
1	158 lbs	0	1,800 lbs	2,500 lbs	39
2	116 lbs	26.5	2,900 lbs	4,000 lbs	38
3	110.7 lbs	30	3,000 lbs	5,000 lbs	67

SEA-CROP® treatment was started at the beginning of year 2.

During this same time period yearly potassium input was also reduced by 7.5% and the phosphorus input remained unchanged.

Significantly, in year 3 the crop came in 2 weeks earlier than others in the area that were planted at the same time. This early harvest characteristic is typical with crops where SEA-CROP® has been applied.

Although this farmer was very successful in substantially reducing nitrogen and potassium inputs in a short period of time, Ambrosia Technology LLC, recommends not cutting those inputs by more than 20% per year. Farms are each different as to their soils, locations and weather so the above results may not be achievable in every case.

In another example, a government agricultural research station in Malawi, Africa grew corn treated with 4 gallons to the acre of SEA-CROP® applied as a soil drench 3 weeks after planting and starter fertilizer (92:21:0+4S). No side dressing was used. The untreated control received both the starter fertilizer and a side dressing of urea so that the control received the full 82 pounds of nitrogen per acre that is recommend for corn production in Malawi. The SEA-CROP® treated test received only 20.5 pounds of nitrogen but gave a crop at harvest that was 97% of the fully fertilized control.

This amounted to a 75% nitrogen input reduction in a single year with only a slight reduction in yield. Again, it is not recommend reducing nitrogen by more than 20% in a single year.

The reader may be wondering at this point if there are any circumstances in which seawater concentrate would fail to give an increased yield and I must, in all honesty, report that there are.

Soybeans often have 25% or more additional pods when treated with SEA-CROP® and this is very beneficial if there is sufficient rainfall later in the growing season so that the pods fill out properly. However, under dry land farming conditions, if rainfall is lacking, the extra pods may turn out to be a liability rather than a blessing and there have been instances of slightly reduced harvests under these conditions.

Another problem has been co-application of SEA-CROP® when mixed with other agricultural products. There are many manufacturers of fertilizers, soil amendments and plant treatment products and these contain a variety of ingredients some of which are compatible with SEA-CROP® and others that are not. For best results, SEA-CROP® should be applied separately unless experience has shown that it definitely is compatible with a given product. Compost and untreated soil amendments such as peat or cottonseed meal present no problems.

In this regard, we must discuss glyphosate aka Roundup® and much space will be given to it in a later chapter. Certain soils such as sandy soils with low humus content, low clay content and low cation exchange capacity, may be so poisoned by glyphosate that seawater concentrate cannot work its magic.

In this chapter we have seen that seawater concentrate is economically important to the farmer apart from any other benefits that it may have such as increased nutrient density or improved soils. In the next chapter we will examine some of its other benefits.

Chapter Four

Nutrient Density

*"Primarily due to discoveries made in the large-scale raising of cattle, hogs, and chickens, we have learned that **trace minerals** are among the most important components of good health—and even life itself. A full complement of the 72-84 trace elements is essential for optimum health. ~ Jon Barron* nutritional researcher

Soils vary widely in their trace mineral nutrient content. When crops are grown and removed from our gardens and farms year after year the soil is depleted of its mineral treasure by some degree each season. The agricultural crops removed from the farm also remove the minerals that they contain. It is a form of mining and it leads to depleted and worn out soils.[1]

It may still be possible to stimulate plants to grow in these soils by adding nitrogen, phosphorus and potassium (NPK) fertilizers, but the nutritional value of these plants will be as incomplete as the soils in which they are grown.

Ultra-trace elements are defined in the scientific literature as essential elements that are required in the daily human diet at a rate of less than 1 ppm (part per million), typically less than 50 micrograms per day (ug/day). A microgram is 1,000,000th of a gram.

The USDA has documented that the trace mineral and ultra-trace mineral content of our worn out farm soils in the USA have undergone very pronounced depletion. As a consequence, the USDA has also documented that the mineral content of food crops and fodder grown on these soils, has undergone severe decline.

Plants grown under optimal conditions may contain as many as 60 different elements. Chemical fertilizers add at best a half a dozen minerals back to the soil each year but that obviously is not doing the job as shown by USDA food analysis data. NPK fertilizer by itself can force abundance but it cannot force depleted soils to make nutritious food.

For example, according to USDA compiled statistics, from 1914 to 1992, a period of eighty years, the mineral content of the average apple has declined severely. As the data in the following table show, calcium has declined by 48%, phosphorus 85%, iron 96%, potassium 2%, and magnesium 83%.

80 Year Decline in Mineral Content of One Medium Apple				
Mineral	**1914**	**1963**	**1992**	**%Change** (1914-1992)
Calcium	13.5 mg	7.0 mg	7.0 mg	48.5%
Phosphorus	45.2 mg	10.0 mg	7.0 mg	84.51%
Iron	4.6 mg	0.3 mg	0.18 mg	96.09%
Potassium	117.0 mg	110.0 mg	115.0 mg	1.71%
Magnesium	28.9 mg	8.0 mg	5.0 mg	82.70%

Source: Lindlarr, 1914; USDA, 1963 and 1997

One might wonder what is the big deal is? The apple still looks good and a lot of apples can be raised by forcing them out of stressed and depleted soils with NPK fertilizers.

Abundance is important to the farmer because he gets paid by the pound, the bushel and the ton, not by mineral content. Abundance is important to the consumer because it helps to keep prices more affordable.

The problem is, according to compiled US Government statistics, as the mineral content of the nation's food declines, so does the health of the populace.

As the following graph shows there apears to be a direct coralation between declining minerals in our nations food supply and per capata deaths by heart disease which were documented to have increased tragically during the same period of time.

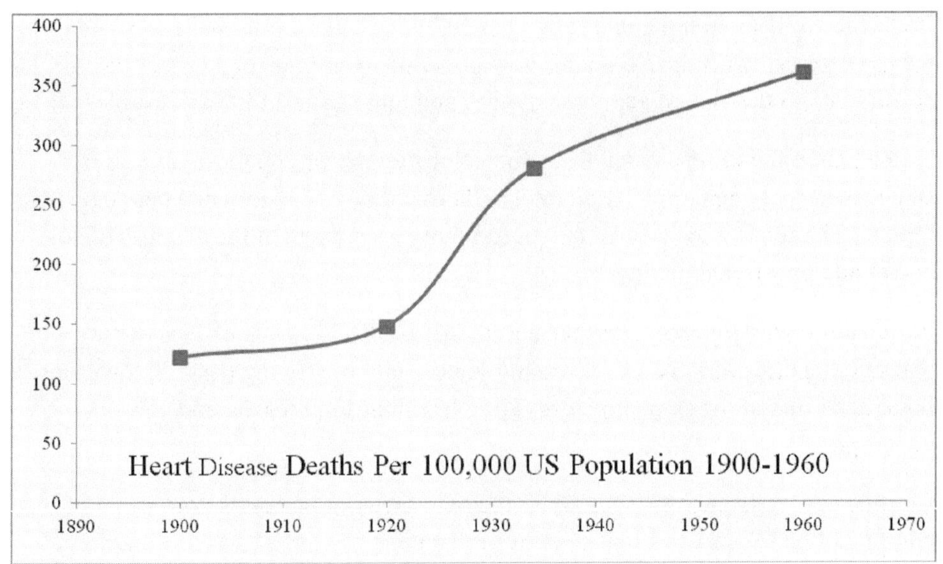

Source: Price, 1938; USDC, 1997

As Minerals Go Down, Disease Goes Up
Changes in the Rates of Selected & Reported Chronic Diseases 1980-1994
Per 100,000 Persons

	1980	1994	% Increase	Mineral Deficiencies Associated with Disease
Heart Conditions	75.40	89.47	18.67	Chromium, Copper, Magnesium, Potassium, Selenium
Chronic Bronchitis	36.10	56.3	55.98	Copper, Iodine, Iron, Magnesium, Selenium, Zinc
Asthma	31.2	58.48	87.44	Magnesium
Tinnitus	22.6	28.24	24.98	Calcium, Magnesium, Zinc
Bone Deformities	84.90	124.70	46.96	Calcium, Copper, Fluoride, Magnesium

Source: USDC, 1996, Werbach, 1993

America's farmers must either make a profit on the farm or quit farming. Since the farmer gets paid by weight and not by nutrient density he has no choice but to follow farming practices proven to produce top notch yields by forcing with NPK.

Bulk trace mineral amendments cost money, money that the farmer must often use for more pressing needs. No matter how good his intentions might otherwise be, he is squeezed in the vice of economic reality and must do what it takes to survive.

Home gardening is very much the same. Nutrition density in most gardens is superseded with high application rates of herbicides, pesticides and chemical NPK inputs. The focus is no weeds, no bugs and big growth, all at the expense of soil health and nutritional quality.

Fortunately, with the seawater concentrate SEA-CROP®, we now have a new tool that will benefit everyone by increasing yield while improving the soil, lowering input costs and at the same time providing healthier food for the end consumer of the crops.

Confined animals must eat what food they are given or perish. Below are more compiled USDA statistics that show what happens when the end consumer is forced to eat the crops that are being harvested in America today.

Changes in Nutrient Content of Beef and Chicken (Per 100 grams)

GROUND BEEF

Nutrient	**1963**	**1992**	**% Change**
Calcium	10.000mg	8.000mg	-20.0
Iron	2.700mg	1.730mg	-35.93
Magnesium	17.000mg	16.000mg	-5.88
Phosphorus	156.000mg	130.000mg	-16.67
Potassium	236.000mg	228.000mg	-3.39
Vitamin A	40.000IU	0.000mg	-100.00
Thiamin	0.080mg	0.038mg	-52.50
Riboflavin	0.160mg	0.151mg	-5.63
Niacin	4.300mg	4.480mg	+4.19

Source: USDA, 1963 and 1997 1993

CHICKEN

Nutrient	1963	1992	% Change
Calcium	12.000mg	10.000mg	-16.67
Iron	1.300mg	1.030mg	-20.77
Magnesium	23.000mg	23.000mg	0.00
Phosphorus	203.000mg	198.000mg-	2.46
Potassium	285.000mg	238.000mg	-16.49
Vitamin A	150.0000IU	45.000IU	-70.00
Thiamine	0.100mg	.069mg	-31.00
Riboflavin	0.120mg	0.134mg	+11.67
Niacin	7.700mg	7.870mg	+2.21

Source: *USDA, 1963 and 1997 1993*

We truly are what we eat or at least our bodies and health are. These statistics only go up to 1992 but there is no reason to believe that the trend established in the first part of the 20th century has done anything but accelerate as we move forward into the 21st.

There is no way to tell by looking at a carrot or tomato if it contains the full range of vitamins and minerals that it would have if it were grown in mineral rich fertile soil.

If all of the minerals that the plant prefers or requires in order to achieve its full genetic and nutrient potential are not present in the soil in available form, the mature plant will have a different physical chemistry and will fall short in nutritive value.

Only chemical analysis can tell the difference between the good the bad and the ugly.

In his book, Sea Energy Agriculture, Dr. Murray recounts two experiments that he performed in the 1950s involving carrots and tomatoes.

In 1958, tomatoes were grown in containers filled with soil. All of the containers received normal fertilization and in addition some of the containers received treatment with sea solids at rates representing 550, 1,100 and 2,000 pounds per acre.

When mature, the tomato fruits were examined for vitamin C content, moisture content and specific gravity. The analytical results that follow show a significant increase in both vitamin C and specific gravity. The increased specific gravity that is observed when crops are treated with seawater concentrate comes from additional minerals taken up by the plant as well as a greater content of metabolites such as additional carbohydrates, vitamins, hormones and proteins.

Sea Solids Tomato Test 1958

Assays	Moisture %	Vitamin C	Specific Gravity
Control	93.7	20.4	0.94
550lbs sea solids	94.2	22.3	1.02
1,100lbs sea solids	93.75	21.2	1.03
2,000lbs sea solids	93.0	25.2	1.11

Sea Energy Agriculture

These results show an absolute relationship between the dosage rate of seawater concentrate and the increase in both vitamin C and specific gravity.

At the top rate of 2,000 pounds of sea solids to the acre, there is a 23.5% increase in vitamin C as compared to the untreated control. The specific gravity, which represents the content of nutritional solids, increased by 18%.

In 1957, carrots were grown in an experimental test plot and treated with sea solids at the rate of 2,200 pounds to the acre. When mature, carrots from the treated test plot together with carrots from the untreated control were sent to a laboratory for analysis of moisture content as well as ash and vitamin A (carotene) content.

The results that follow will come as no surprise to the reader who has seen by now that there is always a benefit when plants are treated with seawater concentrate like SEA-CROP®.

Carrot Test 1957

Assays	Moisture %	Vitamin A	Ash %
Control	90.45	13,000 IU	0.84
Sea solid treated	86.85	18,300 IU	1.36

Sea Energy Agriculture

In regards to vitamin A (carotene) there was a 40.85% increase in the treated carrots and the mineral content, as represented by the ash assay increased by a whopping 61.9%.

Ash is what is left after all of the moisture is evaporated and all of the organic carbon and other organic content is burnt off. Ash represents the mineral cofactors that plants use to build vitamins and enzymes, as well as minerals used for building cell walls, photosynthesis and energy transfer.

The ash contains the mineral cofactors that are the beating hearts of the enzymes that catalyze each and every biological operation within a plant or animal. If trace elements are not available for plants and animals to use as mineral cofactors in building enzymes, the plant or animal can sometimes substitute a different element. There is, however, an efficiency cost when the preferred pathway of enzyme formation cannot be taken. In addition, sometimes a necessary enzyme cannot be made at all due to the missing mineral cofactor. When necessary trace minerals are in scarce supply, plants and animals are unable to achieve their full genetic potential.

The carrot test was repeated again in 1958 with results that, although not as dramatic, are consistent with earlier test results.

Carrot Test 1958

Assays	Moisture %	Vitamin A (Carotene)
Control	89.9	19,800 IU
Sea solid treated	83.3	23,400 IU

Sea Energy Agriculture

The vitamin A content was greater in the treated sample by 18.2% and when subtracted from the moisture contents the compared solids contents show a 65.3% increase in the treated sample.

Tissue analysis of crops treated with SEA-CROP® show very similar results. In 2010 Cornell University personnel performed a study of tomato plants and SEA-CROP® in New York state. Tomato fruit from plants treated with a total of 4 gallons per acre of SEA-CROP® were compared to the fruit of the normally fertilized but untreated control, Vitamin B1 increased by 19%, Lycopene by 97.4% and total carotenoids (Vitamin A) increased by 35.8% over the untreated control.

We have seen from USDA statistics of declining mineral content of food and the US Department of Commerce statistics concerning declining public health that the situation on our nation's farms is far from optimum.

Confined animals have no choice but to eat what is given to them or perish. Due to mineral impoverished soils and liberal and expensive applications of NPK fertilizers, they produce flesh that is marketed as meat for human consumption that has declining nutrient content.[2] The correlation between the demineralized soils, the demineralized crops they produce and declining public health is quite clear except for those who chose to ignore it. Upton Sinclair once said, *"It is difficult to get a man to understand something when his job depends on not understanding it."* This is the current state of our food supply.

What happens when food grown with seawater concentrate is fed to animals and people? Dr. Murray wondered the same thing and conducted tests designed specifically to answer that question.

In 1955, Dr. Murray obtained 306, day old New Hampshire breed chicks. He split these into two equal populations of 153 chicks in each group. One group he fed a diet consisting of two parts corn and one part oats. Both of these grains had been grown on soil treated with sea solids at the application rate of 1,500 pounds per acre. He fed the control group the exact same diet of grains grown on the same farm in the conventional manner in fields not treated with sea solids.
Below are the average weights in ounces, of each chick, weighed at different stages of their growth.

Roosters:	Control Group	Treated Group	% Difference
At 4 months	42 oz.	60 oz.	42.9
At 6 months	106 oz.	128 oz.	20.8
At 2 years	135 oz.	152 oz.	12.6
Hens:			
At 6 months	80 oz.	104 oz.	30.0
At 2 years	96 oz.	114 oz.	12.6
Time of laying	**24.74 weeks**	**21.74 weeks**	
Eggs weight/dozen	19-23 oz.	23 oz.	9.5
Eggs (post 7 months)	24 oz.	28 oz.	16.7
Average feed consumed per pound of weight gained			
Feed Conversion	3lbs	1.86 lbs	
Diseases and Mortality			
Worms	Yes	No	
Nervous disposition	Yes	No	
Leg disjointing	Yes	No	
Mortality	3	0	

Sea Energy Agriculture

Statistics like these should convince us that food we grow for ourselves and our animals, whether in our own gardens or on our farms, has the potential to be the very best and the most nutritious possible. We can guard our health with food that we raise ourselves. It also keeps longer and tastes far better than anything we can buy at the supermarket.

Poultry testing with SEA-CROP® seawater concentrate gave similar results to those obtained by Dr. Murray. During the latter half of 2006 an agricultural feed company in the Netherlands conducted tests with broiler chicks that involved 250,000 individuals.

These animals were supplemented with a minute dose of SEA-CROP® at the rate of 0.02 milligrams per kilograms of bodyweight per day given to them in their drinking water and the following improvements over the untreated controls were observed:

- **Mortality Reduction:**
  In all cases, there was a reduction of mortality in the treated populations as compared to the control groups. Mortality during the vulnerable first two weeks of life was reduced by as much as 90%. Overall average improvement of mortality figures for the entire six week growing cycle was 5% to 15%.
- **Foot Infections:**
  The reduction in foot infections as compared to the control groups was as great as 85%.
- **Skin Abnormalities:**
  The reduction in skin abnormalities was as great as 66%.
- **Weight Gain:**
  There was increased weight gain in the treated animals averaging 3% to 10%.
- **Breast Meat:**
  The ratio of valuable breast meat content was increased by 10%.
- **Feed Conversion Efficiency:**
  There was improvement in feed to meat conversion efficiency as compared to the controls. The range of improvement was 3% to 13%.
- **Health and Temperament:**
  Everyone involved in the testing, from the egg raiser who provided the chicks, to the operator who fed the broiler chicks, to the slaughter house operator stated that they had never seen such a bunch of healthy and calm chickens.

All tests were done at the very low dose of 0.02 ml of SEA-CROP® per kg of bodyweight per day. A dose/benefit study at higher daily dosage has not yet been done.
This experiment indicates that it is the complex organic content of seawater concentrate, working together with the well over 100 minerals and 89 elements it contains, that gives the seawater concentrate its unique, bioactive characteristics.

These results were obtained not by feeding grain grown in treated soil but by treating the animals directly. According to the theories of Dr. Maynard Murray this is not a possibility and it is one of the points where this author's research parts company with his.

As a medical student, Dr. Murray received his education in biology in the 1920s and 1930s. In his book, he states a number of times that plants can only benefit from the inorganic salts in seawater, that they cannot utilize organic matter unless it has been acted upon by bacteria that reduce the minerals it contains to inorganic form. The corollary of this rule is that animals, including bacteria, can only benefit from organic matter as food and not minerals with the exception of salt that Dr. Murray did accept as having a nutritional benefit for animals.

His thinking on this subject was so firm that he stated in the first sentence of his 1963 US *Patent "This invention relates to the process of applying sea water solids as a fertilizer by sea solids we mean the inorganic salts that are dissolved in the water; the term as it will be used in the specification hereinafter does not include living organisms, plant or animal, but means merely the salts that are dissolved, which will include the salts of the various elements mentioned in this application"*[3]

Time marches on and yesterday's scientific rules often become discarded as new findings blur the edges of rigid thought. The Niles Bohr theory of the atom is an excellent example. That theory was good enough to use for building an atomic bomb but it was only partially correct and eventually had to be discarded because it stood in the way of a fuller understanding of reality.

Once I conducted an experiment with a quality bioactive seawater concentrate. It was evaporated to dryness and the remaining salt solids were placed in an oven at 660 degrees Fahrenheit for an hour. This temperature is just high enough to destroy and burn off any organic compounds and organic carbon but will not affect the inorganic content of sea solids. When this material was reconstituted by adding water it was found to no longer exhibit significant bioactivity when applied to plants.

It is fortunate that I had been experimenting with seawater concentrate for half a dozen years and had done experiments with animals before becoming aware of Dr. Murrays pioneering work. Otherwise, I might never have tried supplementing animals directly.

In Appendix A of this book, there is a case study of an experiment with mice performed in 2003 that shows the kinds of dramatic effects that direct nutritional supplementation with seawater concentrates can have.

Another illustration of direct seawater concentrate supplementation has to do with cattle. Dr. Murray mentions in his book, Sea Energy Agriculture, that cattle fed with grain raised on land treated with sea solids were able to gain weight normally, when fed one third less grain than normal.

A similar incident has been related by an organic dairy farmer in the Midwest who maintained his herd through the winter on hay harvested from a field that had been treated with SEA-CROP® seawater concentrate. This man stated that by the end of winter one third of the hay in his barn remained unconsumed. His cattle had thrived through the winter on one third less hay than normal.

These tests produced good results by feeding animals crops grown with seawater concentrate. In comparison is a test of direct supplementation that was performed in the Netherlands by Hak Agro Feed.

Forty veal calves were given a ration of SEA-CROP® added to their feed each day, at the very low dosage rate of 0.02 milliliters per kilogram of body weight. A control herd was given exactly the same amount of feed. When the calves reached the age for slaughter, the group receiving supplementation weighed 8% more than the control group.

Another interesting fact concerning this test is that it was considered a failure because the flesh of the treated veal calves was red instead of the anemia induced grey color that the European consumer has come to expect. The meat was too healthy to be marketed as veal.

One more interesting feature of this test was that it included what are known as goner calves. Once calves are classified as goners, they are expected to die and invariably do. In this case 10 goners were supplemented with a dose of SEA-CROP® on their food each day and they finished the test at the same weight and in equally good health as the rest of the treated population.

In this chapter, we have seen from the work done by Dr. Murray that seawater concentrate, in the form of Mexican sea solids, has a beneficial effect on agricultural crops in terms of both yield and nutrient density. When those crops are fed to animals, the nutritional benefit is transferred onwards.

To achieve these benefits, according to Dr. Murray, sea solids must be applied at the rate of about one ton per acre. The limiting factor in delivering these benefits in more concentrated form is that Mexican sea solids contain 92.64% sodium chloride (sea salt) and only 7.36% other minerals.

By contrast, SEA-CROP® seawater concentrate, when evaporated to dryness, yields solids consisting of 12.3% sodium chloride and 87.7% other marine minerals together with organic substances of marine origin. This is the reason that beneficial effects can be achieved by giving SEA-CROP® directly to animals as a supplement.

By removing 95% of the sodium chloride and the water that it is dissolved in, valuable seawater concentrate containing mostly trace minerals and marine organics is produced. This can then be given directly to animals with benefit. To give a dose of seawater or sea solids containing the same amount of trace minerals we would have to stress the animal with excess sodium chloride.

Before going on to the next chapter and in connection with the theme of this chapter on nutrient density here is a summation of the County Line Farm test data that was presented in chapter three.

The study demonstrated that when six different species of vegetables were treated with a split application of SEA-CROP® at the rate of 4 gallons per acre the following average results were obtained:

Yield increase	+62.56%
Dry matter increase	+12.72%
Brix increase	+35.12%
Protein increase	+19.73%
Insect damage reduction	+52.9%

In the next chapter we shall take a look at the mechanics of how plants grow so that we can see how the seawater concentrate SEA-CROP® affects them in a beneficial way.

Chapter Five

How Plants Grow

"To own a bit of ground, to scratch it with a hoe, to plant seeds and watch their renewal of life - this is the commonest delight of the race, the most satisfactory thing a man can do." ~ Charles Dudley

A plant is a living organism lacking the power of locomotion.

This applies to organisms that range from the largest, the giant redwood *(Sequoia sempervirens)* that grows up to 379 feet tall and can live longer than 3,500 years, to the smallest, the fresh water green algae *(Chlamydomonas),* that is comprised of a single cell less than 1/1000th of an inch in length yet is fully capable of photosynthesis.

An intriguing reason that plants are not capable of locomotion is that they have a non-portable digestive system that is external to their bodies; whereas, animals contain an internal and thus portable digestive system dependent on symbiotic bacteria in their intestines.

Plants are also dependent for their digestion on symbiotic bacteria and fungi that live in the soil encompassed by their root balls, also known as the rizoshpere of the plant. The term **rhizosphere** comes from Ancient Greek: *rhizoma* "mass of roots" and *sphere* "ball". This is a fancy way of saying root ball but it is a useful term which we will be using throughout this chapter.

Botany is the term used for the field of science that studies plant life. In addition to what we normally think of as plant life, the field of botany also encompasses the study of fungi, algae and viruses and includes the study of plant structure, growth, reproduction, metabolism, diseases and chemistry including nutritional and medicinal qualities. Botanists study over a half a million species of plant life.

In this short chapter, we can only skim the surface of a subject of such tremendous complexity, the study of life itself. In the first part of this chapter, we focus on learning a few basic botanical and biological terms so that we can discuss the concepts that aid in our understanding how seawater concentrates can contribute to abundant agriculture.

Enzyme is the most important of terms as no biological operation in either plant or animal life can take place except in the presence of an enzyme. No organism on this planet can exist unless it uses enzymes for each and every one of its life sustaining activities.

The word enzyme was coined in 1881to describe biochemical activity such as takes place in the rising of bread or the fermenting of wine. It was derived from the Greek *enzymes* meaning "enleavened". In connection with this, it is interesting to note that it was only a few years later in 1897that the first crude enzyme extract was isolated unintentionally by the famous German chemist Edward Buchner. By coincidence, and appropriately, the first enzyme extract was isolated from yeast.

Most enzymes are proteins that are biological catalysts that enable and regulate the rate of chemical reactions that take place within an organism. Some enzyme catalysts can speed up reactions a million fold and others can slow them down. In plants, photosynthesis is a catalyzed chemical reaction that takes place in the presence of the enzyme chlorophyll.

Chlorophyll is a name derived from the Greek words *Chloro*s "green" and *phyllon* "leaf". It is a green pigment found in almost all plants. Its job is to intercept photons, energy from solar radiation. It is an enzyme, a protein that catalyzes the transformation of solar energy into carbohydrates. In the heart of each molecule of chlorophyll is an atom of magnesium. Without the mineral magnesium the enzyme chlorophyll cannot be made. Without the enzyme chlorophyll photosynthesis would not take place, plants could not grow and we would all starve. When an enzyme is dependent on an atom of metal for its function, as most enzymes are, the metallic atom is called a cofactor.

A **metallic cofactor** is a metal atom that is bound to, or otherwise associated with, a protein. It is required for the protein's biological activity. These are "helper molecules" that assist in biological transformations that would not take place without their presence. It is estimated that approximately half of all proteins contain a metal.[1]

Now take a deep breath and think about the protein hemoglobin, the red pigment that uses the metallic cofactor iron, to transport the oxygen in that breath to where it is needed, by enzymes that will use it in chemical reactions they will catalyze. If you aren't amazed yet you haven't been paying attention.

Let's take our knowledge of enzyme catalysts and metallic cofactors and move forward to describe some of the activities that plants engage in as they grow.

Those readers who were children growing up as I did in the 1950s will remember looking down a street full of houses where each of them had a television antenna rising above the roof. They looked like the skeletons of metal trees that had lost their leaves in winter. The reason they looked like this is that form follows function and both the television antennas and the parts of trees that are above ground are interceptors of energy radiation.

Trunks, branches, stems and leaves are all part of an antenna array meant to intercept solar energy. It is absolutely correct to think of the above ground parts of plants as solar energy collection and distribution systems.

In green plants, there are two pigment systems, chlorophyll and carotene, that intercept solar photon energy packages and transmute them from energy to mater.

The term **carotene** comes from the Latin word *Carota* "carrot". It is used to describe a group of orange and yellow organic pigments like those in carrots that give their characteristic color to many fruits and vegetables such as carrots, sweet potatoes and cantaloupes. We also see their color in the leaves of deciduous trees as they lose their green color in the fall and present a brilliant display of carotene colors.

Chlorophyll intercepts solar energy in the portion of the electromagnetic spectrum that has a longer wave length than ultra violet light. The carotene pigments intercept energy into the higher ultraviolet portion of the spectrum. Because the carotenes are not a protein and do not contain a metallic cofactor, they cannot process energy themselves so they pass the intercepted energy onwards to chlorophyll by resonance. Carotene also protects chlorophyll from being oxidized and destroyed by ultraviolet light.

It stands to reason that anything that would increase carotene production would help to create a more efficient solar antenna so that a plant could produce more metabolic products through photosynthesis and grow more efficiently. Dr. Maynard Murray documented this increase through tests showing that carrots produced more carotene (vitamin A) when treated with sea solids. Seawater concentrate in the form of SEA-CROP® has been shown to increase the production of total caroteniods in plants. Taken together, although not proof positive, these tests are a good indicators of enhanced photosynthesis and a phenomenon that needs further investigation. It appears that using the seawater concentrate SEA-CROP® for agriculture is like putting rabbit ears on a TV set. Solar energy reception is improved.

Photosynthesis is a term derived from the Greek words *photo* "light" and *synthesis* "process or instrumentality". It is well named and means literally to process with light.

Scientists started studying photosynthesis and unlocking some of its secrets as early as the mid-1600s and steady progress has been made since that time; however, even today the process is only partially understood, and after 350 years of intensive investigation, photosynthesis cannot yet be duplicated in the laboratory.

Basically, the process involves using solar energy to split the water molecule into its separate constituents: Hydrogen and Oxygen. For this operation, plants use an enzyme protein that needs the mineral manganese as a metallic cofactor.

They then combine the Hydrogen and Oxygen from water together with carbon dioxide (CO2), extracted from the atmosphere, to make the simple sugar glucose that becomes the basic building block from which plants construct a wide range of organic chemicals.

So plants take Hydrogen and Oxygen from water that they take up from the earth where their roots are, plus Carbon and Oxygen from carbon dioxide in the air, where their solar antenna rises above the earth. With radiant energy intercepted from the sun's light, they combine these four elements into a simple carbohydrate that is the basic unit from which they then make complex things like sugars, starches, cellulose, fats and proteins.

Photosynthesis is also the source of the carbon in all the organic compounds within a plants body. Photosynthesis is made possible by the enzyme chlorophyll which needs the mineral magnesium as a metallic cofactor.

Plants take part of the energy that is intercepted by their solar antenna and store it in an enzyme that they make for that purpose called *adenosine triphosphate* or **ATP**. This enzyme is one of the most important molecules in all of biology for both plants and animals. It transports chemical energy within cells for metabolism and it is also necessary for cell division. Without it, cells would not have the energy necessary to live and multiply. ATP is a protein enzyme that can only be made with the inclusion of the mineral phosphorus as a metallic cofactor. This is the main reason that phosphorus is a limiting factor in agriculture. If it is not present in soils in available form, ATP cannot be made and plants are unable to transport and utilize energy from their solar antennas in other parts of their bodies.

Complex as this all may sound, remember that even single cell algae and bacteria are able to do all of these processes.

With the use of a multitude of other enzymes, using many metallic minerals as cofactors, plants manufacture amino acids for the formation of proteins, carbohydrates to make glycogen, starch and cellulose. Nucleic acids are formed to build DNA and RNA. Lipids (fats) are made to be part of cell membranes for energy storage. Vitamins are formed to enable certain metabolic reactions. Hormones and chaperone molecules are formed to guide and stabilize growth.

If a plant has light and heat from the sun, carbon dioxide from the atmosphere, water from the earth and a wide variety of minerals with which to form the necessary enzyme catalysts, it can orchestrate all of these processes, grow, flower, fruit and reproduce perfectly, guided only by the hand of God.

It is not surprising that man, in his ignorance and arrogance, has unbalanced these natural processes on a massive scale. Feeding crops synthetic NPK fertilizers, chemical insecticides, herbicides, fungicides, acaricides, algaecides, etc., introduces artificialities and aberrations that distort and pervert the sure inborn knowledge of plants as how to best supply their needs.

This is not to say that mankind cannot be a positive force in agriculture. Of course he can, but it is with a humble awareness of his boundless ignorance that he should consider how to make improvements in what nature has provided both him and the plant to work with.

In the next portion of this chapter we shall take a look below the ground into the strange and wonderful world of the rhizosphere where the sun never shines. Here millions of tiny and microscopic plants, animals and fungi interact with each other, some feeding violently and others feeding symbiotically by exchanging vital nutrients.

Symbiosis is a word from Ancient Greek *sym* (with) and *biosis* "living". It literally means living together and is used to describe a mutually beneficial interaction between different biological species. We will be discussing how plants and animals live together in the soil mutually supporting each other's existence. The media or environment in which this all takes place, below the sunlit world of the atmosphere, is top soil.

What is soil? **Soil** is a mixture of tiny pieces of rocks and the remains of decayed plants and animals. It is formed when rocks are gradually broken up by the weathering action of wind, rainfall and temperature. The expansion and contraction of water in rocks as it freezes and thaws exerts tremendous internal pressures that disintegrate rocks gradually into smaller and smaller particles. Plant and animal debris, left from the life cycle of plants and other organisms growing in the soil, greatly enhance the physical characteristic of the soil know as tilth.

Tilth is a word that comes from the old English word *tilian* and means soil that is ready for planting. It is a term descriptive of both texture and fertility. A soil with good tilth is one that is rich with nutrients and is not compacted with dense sticky clay. Texture is important because oxygen and nitrogen in the atmosphere cannot penetrate into dense compact soil.

A plant's roots secure the plant in soil and, just as the portion of the plant growing above the ground acts as a collection and distribution system, so too does that portion growing below ground. Roots, in addition to anchoring a plant in soil, also seek out and absorb water. Roots are covered with fine hair-like structures called root hairs that are the point of entry for water entering the root collection and distribution system.

Recall that plants need water for photosynthesis and to keep them firm as the bodies of many plants contain up to 95% water. Plants that do not have enough water wilt and droop.

Transpiration is a word used to describe how water enters into roots at the bottom of a plant structure and then travels up through the plant to evaporate out into the atmosphere through pores called stomas, located on the undersides of leaves. As leaves heat up in the sun, the water in the leaves also heats up and then evaporates. The water leaving the plant as water vapor creates a condition much like sucking on a straw, drawing water into the root hairs and then up through the plant.

Plants transpire a lot of water. A large tree might take up through its roots and evaporate out through its stomas 250 gallons in a single summer day.

Groundwater that is taken up into the plant through its rootlets contains dissolved minerals and other nutrients that a plant needs but cannot extract from the atmosphere. Nitrogen is one of these minerals. Although air contains 79% nitrogen, plants cannot use it in elemental form. Some bacteria in the soil can use nitrogen from the air. By utilizing Cobalt, Iron, Molybdenum, Sulfur, and Vanadium as co-factors, Nitrogen-fixing bacteria use an enzyme, such as nitrogenase, to catalyze a reaction that combines nitrogen with hydrogen to form the compound ammonium that is a form of nitrogen plants can use. It is a soluble compound taken up along with many other substances dissolved in the water that a plant eventually transpires.

Nitrogen is required by all life forms including plants for manufacturing proteins especially the all-important enzymes. That is why nitrogen is considered to be a limiting factor in agriculture. In conventional agriculture, when the soil mineral treasure has been expended and soil microflora damaged to the extent that it can no longer generate sufficient nitrogen compounds for the needs of agriculture, the practice has been to apply synthetic nitrogen compounds rather than address destructive practices that caused the problem in the first place.

The long term fix that will deliver a maximum sustainable yield is to encourage proliferation of the diverse flora and fauna that were responsible for the original fertility before the soil was degraded.

To understand how important this concept is we need to have some understanding of the part that soil microflora play in the intricate symbiotic relationships within the rhizosphere of the plant.

During the daytime when a plant is receiving solar radiation, the solar chemical factories in its leaves are busy making up to 100,000 different chemical compounds for its growth and reproductive needs. Of prime importance are the growth hormones: *auxins*, *ethylene*, and *gibberellins*. These are signaling molecules that initiate and regulate cell specialization. They determine which tissues grow upward and which grow downward, leaf formation and stem growth, fruit development and ripening, plant longevity, and even plant death. Hormones are vital to plant growth, and, lacking them, plants would be just a mass of undifferentiated cells.

Another important group of plant growth regulators are the plant cell signaling molecules called *cytokinins*. These are a class of proteins that promote cell division in plant roots and shoots. They are involved primarily in cell growth and differentiation. As more knowledge has been gained the lines have become blurred between the formerly distinct categories of hormones and cytokinins as one has been found to blend into the other. A research agronomist with Cornell University has referred to SEA-CROP® seawater concentrate as having cytokinin-like effects.

During the day, all of the products of plant metabolism are produced and concentrated in the sap of the plant. At the end of the day, when water is no longer evaporating from the stoma on the underside of the plants leaves, the upward pressure on the water column of the plant is relaxed. Sap carrying up to 100,000 chemical compounds, is then exuded back out through the rootlets into the rhizosphere of the plant. At the end of the day, up to 20% of all of the plants photosynthetic products and other chemical manufacturing efforts are expended into the soil.

This is certainly not wasted effort. This nutrient rich root exudate is fed to other life forms such as bacteria and fungi that live in the rhizosphere of the plant. This is not a one-way transaction, but rather a mutually beneficial exchange in which microflora in the soil provide the plant with minerals, vitamins and growth hormones in exchange for nutrients in the root exudates.

Fungi, living in the soil do not have the ability to photosynthesize but some of them do have the ability, through enzymatic action, to extract nitrogen from protein in organic debris. They are happy to trade that nitrogen with a plant in exchange for carbohydrates that the plant produces in its solar factories. A soil with a vigorous fungal population can supply over 50% of a plant's nitrogen needs. Certain bacteria can also fix large quantities of nitrogen from the atmosphere and provide it to plants in exchange for the nutrients in root exudates. These are simple examples of the symbiotic relationships that exist in the rhizosphere of a plant.

A plant is in control of what goes on in its rhizosphere. It can and does create and release in its exudates both antibiotics and growth factors that will eliminate whole classes of bacteria that are not beneficial to it and at the same time encourage population growth of other bacteria that are. In this manner, plants farm the microflora in their rhizospheres, weeding out undesirable species and culturing others, increasing and decreasing their populations according to the changing nutritional needs of the plants as they proceed through their lifecycles.

Plants are also chemical mining engineers. For example, if a plant senses that it needs more iron than has been available, it creates an enzyme called *iron reductase* and exudes the enzyme out through its roots into the soil in its rhizosphere. This process converts iron oxides in the soil into a form that can be utilized by the plant. The chemical composition of root exudates changes constantly as the plant responds to inputs from its environment and then reacts to adjust its environment to its needs.

It should be noted that, just like plants, all of the bacteria and fungi in soil have digestive systems that are external to their bodies. They are, therefore, totally dependent upon enzymes made with metallic cofactors to process food outside their bodies before ingestion.

Mycorrhizal fungi are a class of fungi that live in close symbiotic association with plant roots and they are capable of mining phosphorus from the soil and supplying it to plants in exchange for root exudate nutrients.

Many soils have plenty of the phosphorus that is essential for plant growth but not in a form that is available for plants to use. Phosphorus is a chemical that is tightly bound up in soil and not easily solubilized. Mycorrhizal fungi produce and release *phosphatases*, enzymes that break off phosphate groups from other molecules to which they are attached and then deliver them to the roots of the plant from which they feed. Some fungi are even capable of turning phosphorus into phosphoric acid that is then used to dissolve other valuable minerals out of rock particles.

Most often plants are thought of as only taking up minerals through the root system to feed their foliage. Few are aware that a great deal of the energy captured by the plant through photosynthesis in its leaves is actually used by the plant to manufacture nutrients and other chemicals that it will use to feed its partners in symbiosis underground in the rhizosphere.

The importance of root exudates and multispecies symbiosis in the rhizosphere has only recently come to the awareness of agricultural researchers. It was indeed a matter of "out of sight, out of mind." As a result, these processes are poorly understood. A proper study of root exudates has only just begun and they will be investigated for many years to come.

The bacteria and fungi that feed in the rhizosphere are at the bottom of the food chain. In turn they are eaten by larger microbes such as nematodes and protozoa (amoebae, flagellates and ciliates). These in turn become pray to a multitude of macrofauna of various sizes. From the tiniest bacteria and fungi up to the largest insects and animals that feed on them, it is correct to think of these as packages of fertilizer. As they live and excrete, as they are eaten and excreted, the nutrients that pass through their bodies, or are contained in their bodies, become distributed through the soil so that they become available, ultimately, to plants.

In every step of the way, both in the building up and the tearing down of organic and mineral materials through life processes, enzymes have made each process possible and the entire system relies on minerals to act as active metallic cofactors.

If certain trace elements are not available for plants to use as mineral cofactors in building enzymes, the plant can sometimes substitute a different element. There is, however, an energy cost to the plant when the preferred pathway of enzyme formation cannot be taken. Sometimes a necessary enzyme cannot be made at all due to the missing mineral cofactor.

From the material reviewed in this chapter, we see that presenting a plant with a buffet of 89 elements to choose from, the 89 elements that are detectable in seawater, could make life easier for the plant and for all of the other life forms that live in symbiosis with it. For their life processes, they are all completely dependent upon protein enzymes that use minerals as the active portion of the catalytic molecule. The necessary minerals for the formation of the enzymes must be available to the plant exactly when they are needed or the plant will be set back and possibly stunted permanently.

Experience has shown that the seawater concentrate SEA-CROP® can enable all life forms to more nearly achieve their full genetic potential. If we can help to create the conditions in which soil microflora will thrive, they will do the heavy lifting in providing the plant with its needs. They will also improve both the fertility and the texture of the soil thus giving soil good tilth.

Many subjects have been omitted or only touched upon in a superficial manner in this chapter. Botany and agronomy are subjects too vast to cover in a few pages. However, it is the author's hope that the reader has gained a wider appreciation of the roll that minerals play in life, both above and below the surface of the earth.

Next we shall take a look at how plants respond in a hydroponic environment when seawater concentrate is added to the nutrient broth.

Chapter Six

Hydroponics

"The doctor of the future will no longer treat the human frame with drugs, but rather will cure and prevent disease with nutrition."~ Thomas Edison

Hydroponics is a word compounded from the Greek root words *hydr*o meaning "water" and *ponos* indicating "toil" or "labor". It was coined as a term in 1937 by Dr. W. A. Setchell of the University of California and it was meant to indicate letting water do the work.

Hydroponics is a method of growing plants using mineral nutrient solutions only or in conjunction with an inert substrate for roots to clutch, such as perlite, gravel, mineral wool or coconut husk. This is also referred to as soilless agriculture.

The earliest published work on the subject of growing terrestrial plants without soil was contained in the 1627 book, Sylva Sylvarum by the Elizabethan English philosopher, statesman and scientist Francis Bacon. This was not the first appearance of soilless agriculture as it was reputedly used many centuries before in ancient China and in the fabled Hanging Gardens of Babylon.

The technique was studied by many researchers from the time of Francis Bacon forward and led directly to the idea that plants only need a handful of nutrients to grow. Modern, large scale hydroponic greenhouse operations grow plants in solutions containing 12 elements: nitrogen, phosphorus potassium, calcium, magnesium, sulfur, iron, manganese, zinc, boron, copper and molybdenum.

While this might be fine for growing orchids and other flowers to look at, this technique will produce food crops containing only the elements that were available in the nutrient solution. If you wonder why this might be a problem, think back to the last hothouse tomato that you tasted.

The basic concept of soilless agriculture is that in nature, soil acts as a mineral nutrient reservoir but the soil itself is not essential to plant growth. Instead of bringing water to the plant grown in soil, the plant can be brought to the water and raised under closely controlled conditions where it is liberated from the natural conditions of drought and insect pests.

When done under glass, as in a greenhouse, hydroponics does have the additional potential advantages of consistently high yields and growth of plants out of season or in cooler climates than would be possible with in-ground agriculture. Another consideration is that in some commercial installations, in a well controlled environment, it is possible to grow plants without insecticides and of course weeding and herbicides are not needed.

A century ago, it was thought that plants only needed a handful of elements to thrive and produce nutritious food. The importance of microflora symbiosis in the rhizosphere was unknown and little was known about mineral cofactors for enzyme formation in both plants and the animals that eat them.

This author believes that a person eating only conventional hydroponically raised food would starve to death in a short time no matter how much of it they ate. If they didn't starve, they would probably die of heart disease or cancer. It is simplistic to think that properly nutritious food can be grown with only a handful of elements.

The study of hydroponics became popular during the 1930s and Dr. Maynard Murray became interested in the subject. As a result, he spent many years doing hydroponic experimentation and in his later years, even had a commercial hydroponic operation in Florida. There he grew tomatoes using sea solids added to water as the mineral source for his nutrient supply.

Through decades of experimentation, he developed a superior method of soilless agriculture. He supplied all of the minerals available in seawater to plants in the form of diluted solutions of seawater or sea solids. This allowed plants to choose from a complete banquet of all the soluble marine minerals, provided in the correct proportions for the plant's best use. Other than seawater minerals, the only addition to the nutrient solution was nitrogen, in a plant available form, so that it could be used by plants for building proteins. This combination allowed plants unrestricted fulfillment of their mineral needs so that the all-important enzymes and cytokinins could be formed as needed by the most efficient pathway.

One experiment in particular proved conclusively that plants grown with sea minerals, in addition to tasting much better, were far healthier than those grown by the hydroponic methods prevailing at that time.

Dr. Murray filled a cement trough 100 feet long by 3 feet wide and 8 inches deep with sterilized marbles for roots to clutch, in place of soil. Tomatoes were planted in this medium and were watered three times each day with 5,000 gallons of water containing 112 pounds of sea solids dissolved in the solution.

A second similar hydroponic bed was planted in the same manner but was watered three times each day with a traditional hydroponic mineral nutrient solution.

He devised a dramatic test to show comparative health and disease resistance between the two plant populations. All of the tomato plants in the trial, both the seawater treated and the untreated control populations were sprayed with a solution containing Tomato Mosaic Virus, a plant disease that is lethal to tomatoes.

As a result of the viral inoculation, all the plants in the control population succumbed to the virus and did indeed die of Tobacco Mosaic Virus. Interestingly, the seawater treated plants did not contract the disease and remained healthy.[1]

Unfortunately, Doctor Murray's pioneering research was not adapted by the commercial hydroponic industry and today the situation within that industry is little different than it was decades ago. Plants are still being raised in nutrient broths containing seldom more than a dozen elements. As a consequence, the foods produced by soilless agriculture are at best equal to the poorest crops forced to grow on completely worn out soils with heavy applications of NPK fertilizer.

Fortunately, a new player has arrived on the scene in the form of SEA-CROP® ocean mineral concentrate. It is an easier way to provide seawater minerals for hydroponics and because 95% of the sodium chloride contained in seawater has been removed from the concentrate, industry is far more receptive to its use.

The Dutch are the recognized world leaders in commercial hydroponics. The Netherlands has a total hydroponic production area of some 25,000 acres under glass, made up of 13,000, mostly family-based operations that employ an estimated 40,000 people.

Nearly all greenhouse production in the Netherlands is hydroponic. Fifty percent of the value of all fruit and vegetables produced in the country comes from hydroponic production. The Netherlands' most important vegetable crops are peppers, tomatoes and cucumbers with much of this production being exported to other countries in Northern Europe. Roses, daisies, carnations and chrysanthemums for cut flowers are also grown hydroponically primarily for export markets.

Since 2005 Hak Agro Feed, an 80 year old Dutch company, has been distributing SEA-CROP® in Northern Europe for agriculture and animal nutrition under their brand name *Immutines*.

They have conducted numerous field trials with in-ground and hydroponic crops. Hak Agro Feed has also conducted animal trials with both broiler chicks and cattle. In this chapter we are discussing hydroponics so we will next look at some of the very pertinent work Hak Agro Feed did investigating dose/benefit relationships in regards to improved yields and antioxidant content in hydroponically grown foods.

Hak Agro Feed documented benefits from the addition of small amounts of SEA-CROP® to hydroponic solutions: earlier ripening, increased yield, increased mineral content, improved storage life and increased ORAC.

What is **ORAC**? It is an acronym standing for Oxygen Radical Absorbance Capacity which is a test that quantifies general plant health and nutrient levels. More specifically, it quantifies antioxidant activity similar to the way brix testing measures the total dissolved solid content in plant sap. ORAC readings indicate general plant health, pest and disease resistance and, most importantly, quality.

Most people have heard of important antioxidants such as vitamin C and vitamin E but plants make many antioxidants and the ORAC test measures the combined total protective power of all a plant's antioxidants taken together.

Plant ORAC Test Results with SEA-CROP® in Hydroponic Production

Testing has proven that adding SEA-CROP® to a hydroponic nutrient solution enhances ORAC activity significantly. There have also been very positive impacts on both yields and keeping quality. In SEA-CROP® treated cucumbers, ORAC activity increased 37% and the yield increased 7.1%. In SEA-CROP® treated tomatoes, ORAC activity increased 13.6% and the yield increased 31.4%.

In the Appendix of this book are English translations of a few of the ORAC testing reports issued to Hak Agro Feed from the testing laboratory ORAC Europe BV.

The following text is an English translation of part of a report written by Doctor E. van den Worm, of ORAC Europe, for Hak Agro Feed. It describes the importance of antioxidants for plant and animal health and their place in human nutrition. Some of Doctor van den Worm's experiences testing SEA-CROP® are also summarized. This paper helps provide more information about ORAC and the effect SEA-CROP® has on plant ORAC activity.

Antioxidants and Free Radicals

Antioxidants are molecules that are capable of protecting the human body against so-called 'free radicals' and other harmful compounds that are released within the body as a result of oxidative reactions. Free radicals are very reactive oxygen metabolites which, when produced in excess amounts, can cause damage to tissues and organs. Common free radicals are for example superoxide anion (O_2), hydroxyl radical (·OH) and lipid peroxyl radical (ROO·).

Antioxidants are able to neutralize free radicals by accepting the unpaired electron that makes free radicals so highly reactive. Scientific research has shown that antioxidants can play a protective role in a number of diseases in which oxidative stress and free radicals play a role such as cardiovascular and neurodegenerative diseases.

In humans, antioxidants can originate from external (food) sources as well as being produced endogenously. Well-known examples of antioxidants originating from food are vitamins such as ascorbic acid (vitamin C) and a-tocopherol (vitamin E), but also a number of plant-derived compounds possess strong antioxidative capacities. An important group of bioactive molecules showing potent antioxidative capacities are plant constituents such as anthocyanins (plant-pigments), carotenoids and polyphenols (especially flavonoids) from vegetables and fruits and fat-soluble antioxidants from vegetable oils.

Plants are capable of producing numerous amounts of very diverse molecules. Besides molecules originating from the so-called primary metabolism which mainly serve processes such as CO_2 assimilation, energy exchange and growth, plants contain many molecules which are products of the secondary metabolism. These secondary metabolites are all more or less produced to protect and maintain the plant itself. Some molecules fend off predators like insects and herbivores, while other molecules attract insects that are important for pollination. The amounts of secondary metabolites within a certain plant are very variable and are highly dependent on all sorts of (environmental) factors. Thus, plants that are under attack by insects or fungi will contain higher concentrations of protective molecules than plants that are free of attacks. Also external factors such as soil conditions, soil hydrology and sunlight conditions can cause significant differences in (secondary) metabolite concentrations. Furthermore, in commercially important plants such as vegetables and fruits, specific varieties and cultivars also play an important role.

Plant-derived Antioxidants

Plants can produce a huge variety of antioxidative molecules. Anthocyanins and pigments, responsible for many of the bright colors of fruits and flowers, offer the plant protection from free radicals originating from oxidative stress or exposure to UV radiation.

Flavonoids form another important group of molecules that possess strong antioxidant capacity and thereby are able to protect the plant from all kind of potential harmful oxidation processes. Because many antioxidants also belong to the secondary plant metabolites, the amounts of these molecules in plants can also significantly fluctuate depending on plant species and environmental conditions. Concentrations of plant antioxidants are also highly depend on the developmental stage of the plant.

The fact that plants can react to outside threats by producing protective molecules may very well explain some of the recent observations by several scientific study-groups investigating the quality of organic vegetables and fruits. In press releases by leaders of this European research, it was revealed that organic crops show higher amounts of antioxidants than conventionally grown crops. It was shown that organic tomatoes, wheat, potatoes, cabbages and onions contain 20-40% more antioxidants. They also showed that organic milk contained on average about 60% more antioxidants, among them, 50% more vitamin E and 75% more carotene. These findings seem to confirm the idea that plants in fact do not just 'randomly' produce all kind of constituents, but that these metabolites play a vital role in the survival strategy of the plant. After all, in organic farming no or far less pesticides are used. This may give insects and fungi a bigger chance to attack these plants. That will lead to enhanced stress levels in these plants. Therefore, these plants have to produce higher concentrations of protective molecules to maintain themselves.

So, by exposing plants to increased stress levels, these plants may be stimulated to produce higher amounts of secondary metabolites, thereby also producing higher concentrations of antioxidants.

Effect of SEA-CROP® on antioxidant capacity of cucumbers

In a total of three studies, effects of SEA-CROP® addition to the water during the growing process of cucumbers were investigated. Cucumbers were rooted on Rockwool and grown in greenhouses and during distinct periods, several concentrations of SEA-CROP® were added to the water which was administrated to the growing plants (0.5, 1, 1.5, 3and 4 ml/m2/week, respectively).

From data delivered by the growing company, it appeared that after harvesting the average weight of the SEA-CROP®-treated cucumbers was slightly higher than that of the control cucumbers (non-treated cucumbers).

Additionally, a mineral analysis showed that the SEA-CROP®-treated cucumbers absorbed 9-20% more of the administrated mineral mixture compared to the control cucumbers.

By using the standardized and validated ORAC (Oxygen Radical Absorbance Capacity) method, the antioxidant capacity of the SEA-CROP®-treated cucumbers was determined and compared to that of the control cucumbers (non-treated cucumbers).
(Also see ORAC Europe reports of September 2008, May 2009 and July 2009).

From the results obtained by the antioxidant capacity determinations performed by ORAC Europe BV, the following conclusions can be drawn:

Antioxidant capacity:

The antioxidant capacity of SEA-CROP®-treated cucumbers (expressed as hydrophilic ORAC value) was 29-37% higher than the antioxidant capacity of control cucumbers.

It appears that addition of SEA-CROP® to the water during the growth process may lead to a higher concentration of antioxidants or molecules with an antioxidative effect in these cucumbers, which ultimately results in a higher ORAC value (per unit of weight). However,no concentration-dependent effects of SEA-CROP® administration were detected. The measured increases in antioxidant capacity after administration of SEA-CROP® in concentrations of 0.5, 1, 1.5, 3 and 4 ml/m2/week, were 8.7%, 36.6%, 33.3%,37% and 29.7%, respectively.

The increase of antioxidant capacity already seems to reach a maximum level after administration of a SEA-CROP® concentration of 1 ml/m2/week (and higher). Probably, further increasing the SEA-CROP® concentration will not lead to a further increase in antioxidant capacity.

This paper by Dr. van den Worm summarizes a few of the findings from three years of testing the results of SEA-CROP® additions to conventional 12 element hydroponic solutions under commercial greenhouse conditions.

SEA-CROP® was added to the nutrient solutions on a weekly basis and this was done in trials with rates of additions that varied from ½ to 5 milliliters of SEA-CROP® per square meter of greenhouse. These are minute amounts.

As stated by Dr. van den Worm, plant production of ORAC response appeared to be maximized once the seawater concentrate was added to the nutrient solution at the rate of at least one milliliter per week for each square meter of greenhouse space. Additions at the rates of 2, 3 and 4 milliliters per square meter per week gave ORAC test results that were essentially the same.

It was very much a different story with regard to yield increases. There was a dose related improvement in yields with each incremental addition of seawater concentrate. The sweet spot for increased yields in relation to the amount of product added to the nutrient solution appeared to be between 3 to 4 milliliters added per square meter of greenhouse space per week.

It is interesting that addition of SEA-CROP® enabled plants to uptake 20% more of all the other conventionally used minerals contained in the nutrient solution. This was measured by assaying the nutrient solutions and monitoring additions of minerals used to refresh the solution.

The enhanced mineral uptake also showed up in the increased weights of dry matter content. This of course was also, at least in part, responsible for improved shelf life. Even at the very low application rate of ½ milliliter per square meter, shelf life was noticed to have significantly improved.

Again this demonstrates that it is not just a matter of opinion that conventionally grown foods are nutritionally deficient in comparison to those that can be grown with the use of the seawater concentrate SEA-CROP®. This is as true of soilless agriculture as it is of crops grown in God's good earth.

Chapter Seven

Which Seawater Concentrate is Best

"Excellence is never an accident. It is always the result of high intention, sincere effort, and intelligent execution; it represents the wise choice of many alternatives - choice, not chance, determines your destiny." ~ Aristotle

There are a number of seawater concentrates on the market today. Ambrosia Technology, LLC believes its product, SEA-CROP®, is superior. That belief is not arbitrary as there are a number of reasons derived from science and data from field trials to substantiate this belief.

Beginning with science, Ambrosia Technology, LLC has put a significant investment into both spectrographic and chemical analysis of its own and competing seawater concentrate products in order to guide its research. We will take a look at how the contents of various types of seawater concentrates compare as assayed by Inductively Coupled Plasma - Mass Spectroscopy (ICP-MS).

There are four types of seawater concentrate in the market place. The first is a dry product that will be referred to in this chapter as "Sea Solids". Next is a liquid seawater concentrate made in Florida and that product shall be referred to as "Ocean Minerals". A third product that comes from Australia appears to be a byproduct of the solar salt industry. We shall refer to it as "Aussie Bittern". The fourth product is SEA-CROP® made by Ambrosia Technology LLC in the State of Washington at the mouth of Willapa Bay.

The existence of four products validates market acceptance of seawater concentrates. From the documented trials in this book and the details below, we demonstrate hands down that SEA-CROP is the best value and superior in all categories.

Every naturally occurring chemical element on earth is to be found dissolved in seawater and, except for the element Phosphorus, they are always in relatively constant ratio in proportion to one another. These elements in solution are all in salt form. For example, the element Sodium is mostly in the form of sodium chloride.

When in solution, the two parts of a mineral salt molecule dissociate so that sodium chloride, for example, would no longer have a crystal relationship but would be more loosely bound, separate ions of sodium and chlorine each with their own electrical charge. In this condition such a compound is called a crystallite because if the solution were to be evaporated to dryness the compound would form crystals.

Seawater has a dissolved mineral content of approximately 3.47% and this for the most part is constituted of six ions.

Element	**Symbol**	**% Dissolved Inorganic Content of Seawater**
Chlorine	Cl	55.04%
Sodium	Na	30.61%
Sulfate	SO4	7.68%
Magnesium	Mg	3.69%
Calcium	Ca	1.16%
Potassium	K	1.10%
Total		**99.28%**

In addition four more minor ions make up much of the other 0.72% as follows:

Element	**Symbol**	**% Dissolved Inorganic Content of Seawater**
Bromine	Br	0.19%
Strontium	Sr	0.07%
Boron	B	0.04%
Fluorine	F	0.001%

Together, these ten ions comprise 99.541% of the dissolved inorganic content of seawater with most of the remaining 0.419% made up of dissolved carbon dioxide gas.

As can be seen, the trace and ultra-trace minerals are scarce even in seawater. If the object is to supply plants with mineral nutrition containing as much of the marine trace and ultra-trace minerals as possible, then the most desirable seawater concentrate would be one that efficiently concentrates the trace minerals without diluting them with sodium chloride.

Sea Solids

Sea Solids come from a deposit of salt flats located on the West side of the Gulf of California, also known as the Sea of Cortez. About 125 miles South of El Centro, California and the Mexican border town of Mexicali lies the small resort town of San Felipe. About 15 miles north of town there is a solar salt works where sea salt is made. Starting there and extending much of the way north to Mexicali are salt flats in an area that has in the past been repeatedly inundated by seawater. This inundation partially evaporated and left behind a portion of its mineral content. Although this material is mostly sea salt, it does have some trace mineral content and also a moderate degree of bioactivity when applied in sufficient quantity. Dr. Murray named this salt material "sea solids".

Let's take a look at how the major ions in sea solids compare to a salt made by completely evaporating seawater while retaining all of its mineral content.

Element		Complete Sea Salts	Sea Solids
Chlorine	Cl	55.04%	56.644%
Sodium	Na	30.61%	36.444%
Sulfate	SO4	7.68%	5.4%
Magnesium	Mg	3.69%	0.3%
Calcium	Ca	1.16%	0.626%
Potassium	K	1.10%	0.109%
Bromine	Br	0.19%	0.0129%
Strontium	Sr	0.07%	0.0099%

This comparison shows that Sea Solids contain more sodium chloride and less of the other minerals than were contained in the original seawater. The reason for this is that when seawater evaporates, eventually the solution becomes saturated with mineral content and can hold no more. At that point, crystals begin to form and fall out of solution. Because sodium chloride is less soluble and in greater concentration than the other minerals, it is the first to crystallize out of solution.

This is the method used in salt works for making solar salt. After the sodium chloride crystallizes out, the remaining mineral liquor, called bitterns, is drained off and either retained for further processing or discarded. The solar salt, mostly sodium chloride, can then be scooped up for drying and packaging.

Notice in the comparison that the Sea Solids contain less than 10% of the magnesium content that complete dried sea mineral salts contain.

Bitterns contain a large percentage of magnesium chloride that is very difficult to dry out because it is a deliquescing mineral. That means that it draws moisture out of the atmosphere to satisfy its thirst. Because of this characteristic, bitterns, which contain most of seawater's trace and ultra-trace minerals, tend to separate out from salt flat deposits such as those on the Baja, Mexico salt flats where sea Solids originate. The bitterns either sink into the ground when the water table lowers or are rinsed out during the next flooding cycle.

Dr. Maynard Murray recommended application of sea solids at the rate of 1,500 pounds to the acre. In Ambrosia Technology LLC's opinion, such a high application rate is necessary for this form of seawater concentrate because the bioactive trace and ultra-trace minerals are in limited supply in relation to the sodium chloride content.

It is possible to add such huge amounts of sodium chloride to agricultural soils under some climate conditions without causing damage to the land or the crops grown in them. But that cannot be said for all soils and all climates.

When such large amounts of sodium are added to agricultural soils with low cation exchange capacity (CEC), they may become saturated with the sodium ion. Then the ability of these soils to absorb and exchange other minerals would become compromised to the determent of crops grown in them.

Also, many soils under irrigated conditions suffer from salt build-up and crops suffer as a consequence. This condition can become so severe that the soil becomes sterilized and infertile. It is a condition that should not be aggravated with large applications of sodium chloride.

Ocean Minerals

The next form of seawater concentrate is, Ocean Minerals. This is a seawater concentrate that appears to be made by reverse osmosis. It is unknown to this author if the concentrate is made directly for use in agriculture or if it is a byproduct from seawater desalination for drinking water. This is a complete mineral concentrate that preserves all of the minerals contained in the original seawater and, furthermore, it contains much of the very important bioactive organic content.

The processing method used to make Ocean Minerals does not concentrate all of the minerals equally. Depending on which element we look at, the minerals have been concentrated from 2.5 to 4 times the mineral content of Seawater and the sodium chloride is retained.

In side by side trials, when the Ocean Minerals product was applied to plants at the same application rate as SEA-CROP®, the product that always gave, by far the best results, was SEA-CROP®.

These side by side trial results are probably do to the fact that Ocean Minerals is much more dilute than SEA-CROP® and contains mostly sodium chloride.

SEA-CROP ®

SEA-CROP ® is a unique product made by a proprietary process that removes 95% of the sodium chloride while concentrating the other minerals between 2,000% to 3,000%. The degrees of concentration depending upon which mineral as they do not all concentrate evenly.

The product has been many years in development and has gone through a number of upgrades since it was first introduced into the market place.

Research is ongoing and Ambrosia Technology, LLC constantly strives to produce the most bioactive seawater concentrate available anywhere in the world and to make it available to the public at the most competitive price.

The table starting on the next page compares the mineral content of SEA-CROP®, Ocean Minerals and Sea Solids.

These assay results were obtained by Inductively Coupled Plasma - Mass Spectroscopy (ICP-MS) assays performed by Acme Labs of Vancouver, BC, Canada.

The letters BLD stand for Below the Limits of Detection. Note that for both Ocean Minerals and Sea Solids, many of the elements, if present, are at levels below the limits of detection.

PRODUCT COMPARISON ASSAYS			
	SEA-CROP®	**OCEAN MINERALS**	**SEA SOLIDS**
Element	**PPM**	**PPM**	**PPM**
Aluminum, Al	0.265	BLD	BLD
Antimony, Sb	0.127	BLD	BLD
Arsenic, As	0.328	BLD	BLD
Barium, Ba	0.128	BLD	0.61
Beryllium, Be	0.056	BLD	BLD
Bismuth, Bi	0.00002	BLD	BLD
Boron, B	247.892	7.0218	BLD
Bromide, Br	755.203	170.883	192.36
Calcium, Ca	475.167	1,361.01	6,263.97
Cerium, Ce	0.00156	BLD	BLD
Cesium, Cs	0.002	BLD	BLD
Chloride, Cl	74,456.0	43,774.0	5,664,400.0
Chromium, Cr	0.237	BLD	BLD
Cobalt, Co	0.113	BLD	BLD
Copper, Cu	0.54	BLD	BLD
Dysprosium, Dy	0.001	BLD	BLD
Germanium, Ge	0.00896	BLD	BLD
Indium, In	0.11	BLD	BLD
Iron, Fe	15.183	21.559	167.46
Lead, Pb	0.113	BLD	BLD
Lithium, Li	28.21	0.57	2.18
Magnesium, Mg	32,701.2	4,158.54	3,503.65
Manganese, Mn	0.933	BLD	6.53
Molybdenum, Mo	0.157	BLD	BLD
Neodymium, Nd	0.036	BLD	BLD
Nickel, Ni	0.277	BLD	BLD
Phosphorus, P	16.047	BLD	BLD
Potassium, K	2,857.447	1,245.283	1,091.62
Rhenium, Re	0.0037	BLD	BLD
Rubidium, Rb	0.433	0.346	0.3
Selenium, Se	5.231	BLD	BLD
Silicon, Si	10.377	BLD	BLD
Silver, Ag	0.153	BLD	BLD

Sodium, Na	8,445.33	29, 815.69	364,437.0
Strontium, Sr	5.125	22.165	9.92
Sulfur, S	5,497.0	4,208.0	18,160.0
Tungsten, W	0.065	BLD	BLD
Uranium, U	0.019	BLD	BLD
Vanadium, V	1.184	0.346	3.95
Zinc, Zn	1.31	1.372	5.51
Zirconium, Zr	0.057	BLD	BLD

ASSAY NOTES

1. The assay values are reported as parts per million.

2. The assay values reported for SEA-CROP®, "Ocean Minerals" and "Sea Solids" are from ICP-MS assays performed on those products at Acme Labs of Vancouver, BC, Canada.

3. The Letters **BLD** indicate that element, if present, is below the limit of detection by the assay method used.

4. Note that some elements such as the very important trace minerals Cobalt, Copper, Phosphorus and Selenium are present in SEA-CROP® at levels well above the limits of detection but not in the other two products. Also note that the other two products contain between 3 to 43 times more sodium than SEA-CROP®, 8 to 9 times less crucially important magnesium and less than half the essential potassium.

5. Although seawater concentrates are documented to contain 89 elements in measurable amounts, all of the elements cannot be measured by any single assay method. The 41 elements in the table above are those elements detectable as present in seawater concentrate by ICP-MS.

6. Of the 41 elements examined, SEA-CROP® contains higher levels than Ocean Minerals for 36 elements or 87.8% of the categories examined. SEA-CROP® contains higher levels than Sea Solids for 32 elements or 78% of the categories examined.
   Both of the other products contain much more of the potentially harmful sodium ion.

SODIUM CHLORIDE COMPARISON--WHY PAY FOR SALT?

1. SEA SOLIDS is a dry product containing 92.64% sodium chloride by weight.

2. OCEAN MINERALS is a liquid product.

   When dried it yields 7.5% solids consisting of 77.81% sodium chloride and 22.19% other minerals.

3. SEA-CROP® is a liquid product.

   When dried it yields 20.0% solids consisting of 12.3% sodium chloride and 87.7% other marine minerals and organic substances of marine origin.

BITTERNS

The final category of seawater concentrate to consider is bitterns. One would think that bitterns would be very bioactive because of all the trace and ultra-trace elements they contain, but Ambrosia Technology, LLC has not found that to be the case. Bitterns byproducts from three different solar salt operations, including Australian bitterns, were trialed against SEA-CROP® and in each case SEA-CROP® outperformed the bitterns by considerable margins with regard to plant growth response.

It may be that exposure to ultraviolet light, high temperatures and increasing salinity during the lengthy solar evaporation process degrades and alters the organic content of bitterns in a way that diminishes and destroys its efficacy.

Although an ICP-MS assay was not performed on "Aussie Bittern", it was examined for sodium content and found to contain 249% more sodium ion than SEA-CROP®.

SAFE

SEA-CROP® has also been tested for yeast, mold, colony forming bacteria, total heavy metals and organic arsenic. It has been given a clean bill of health on all counts.

Chapter Eight

How to Use It

"There are no such things as applied sciences, only applications of science."

~ Louis Pasteur

As with all seawater concentrates, SEA-CROP® seawater mineral concentrate is nontoxic and very safe to use. It has been approved by the US FDA for use as a mineral supplement in animal nutrition. One need not worry that people, pets or farm animals will come to harm from the presence of SEA-CROP® in their environment.

Thus far, all animal and plant species tested, without exception, have benefited from the application of SEA-CROP® seawater mineral concentrate.

SEA-CROP® is very versatile and can be used in a variety of ways. Some large corn and soybean farmers have good results with a single application each season. This can take place either in the furrow with the seed, at planting, or after emergence when the plants are 6 to 8 inches high.

There are truck farmers with high value crops who use SEA-CROP® as a foliar spray every two weeks. Trials using the product as a root dip or seed soak have given good results even when no further follow-up treatment was given.

Following are the procedures that Ambrosia Technology, LLC believes will give the best results for general use.

SEA-CROP® seawater extract may be used on all food and non-food crops. It may be applied directly to the soil, to the roots or applied as a foliar spray.

Following are the general instructions for use of SEA-CROP® seawater concentrate.

SEA-CROP® Instructions for Use

- **Dilution:**
  The product must be diluted before application.
- **Concentration**
  Use at a concentration of 1% to 2% strength. One gallon of SEA-CROP® added to 49 gallons of water equals a 2% solution. Five tablespoons of SEA-CROP® (two and a half ounces) per gallon of water equals a 2% solution.

- **Soil Drench:**
  Diluted SEA-CROP® can be used as a soil drench. Use a minimum of 3 applications per season applied at 3 week intervals starting at planting, transplanting or after emergence. A good alternative is to do one soil application followed by two foliar applications after emergence.

- **Foliar Spray:**
  Diluted SEA-CROP® can be used as a foliar spray. A minimum of 3 applications per season applied at 3 week intervals are recommended.

Annual Application Rates

Garden Produce:

- Apply 4 gallons of SEA-CROP® concentrate per acre.

Row Crops:

- Apply 2 to 4 gallons of SEA-CROP® concentrate per acre.

Trees and Orchards:

- Medium size trees (size 3-6 feet): use 4 oz. SEA-CROP® seawater concentrate per tree, not to exceed 10 gallons per acre.

- Large trees (size 6-12 feet): use 6 oz of SEA-CROP® concentrate per tree, not to exceed 10 gallons per acre.

Lawns and Turf:

- Apply 2 to 4 gallons per acre.

Safety:

- Safely handle SEA-CROP® as you would any agricultural fertilizer.

For best results, never mix SEA-CROP® with synthetic fertilizers, insecticides or herbicides. SEA-CROP® should be applied, by itself, one to two weeks before or after other agricultural treatments. Co-application with compost tea is permissible.

Excellent results have been reported when SEA-CROP® has been mixed with molasses or unpasteurized milk for co-application. Either of these is an effective addition to aid in building up soil microflora populations and promotes rapid assimilation by soil biota of all kinds.

When using molasses, any kind will do. Just pick up the cheapest kind that you can get at the feed store. Also take care not to overdo it. If you use more than a couple of gallons per acre per year, you run the risk of turning soil bacteria into sugar junkies and that is counterproductive.

When milk is used, 2 to 3 gallons of unpasteurized milk will do the trick. Alternatively, other sugar sources such as 6 pounds to the acre of sucrose or dextrose could be used.

A good program is to only use molasses or milk with the first SEA-CROP® application each growing season.

Seed Soak:

- SEA-CROP® may be used as a seed treatment. For each acre to be planted, dilute 1/2 pint (8 oz.) of concentrate with 24 pints of water and soak the seed to be planted overnight prior to planting. This will result in superior germination and vigorous seedlings.

Animals:

- SEA-CROP® may be used as a mineral supplement for animal nutrition. The product can be added to drinking water or food. The very low daily dosage rate is 0.02 to 0.05 milliliters per kilogram of body weight. This is approximately 1/5 to1/2 teaspoon per 100lbs.

In the next chapter we learn that some modern agricultural practices are unwholesome and destructive. Seawater concentrates can help the conventional farmer to wean himself from dependence on chemicals that poison soil and lower its fertility.

Increased crop yields obtained at the University of Louisiana at Baton Rouge.

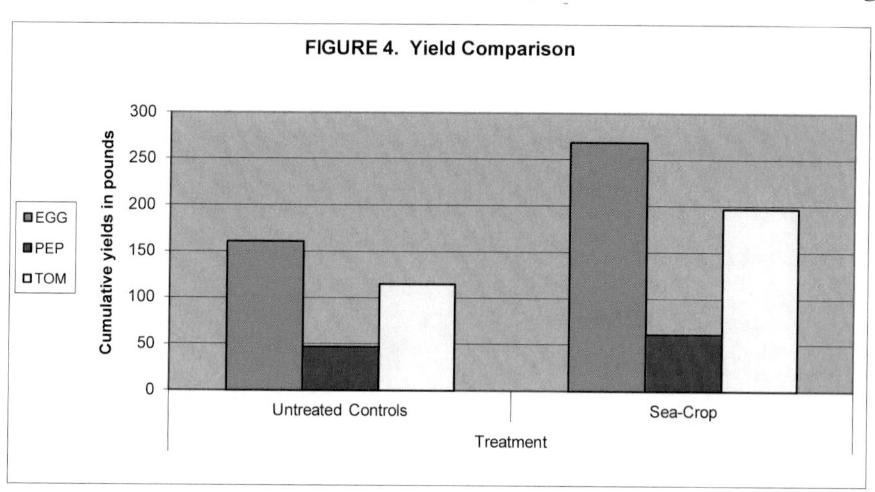

FIGURE 4. Cumulative Eggplant (EGG), Pepper (PEP) and Tomato (TOM) Yields (in pounds) Across 6 Harvests for 2008 Cal-Agri Trial as Influenced by the Treatment .

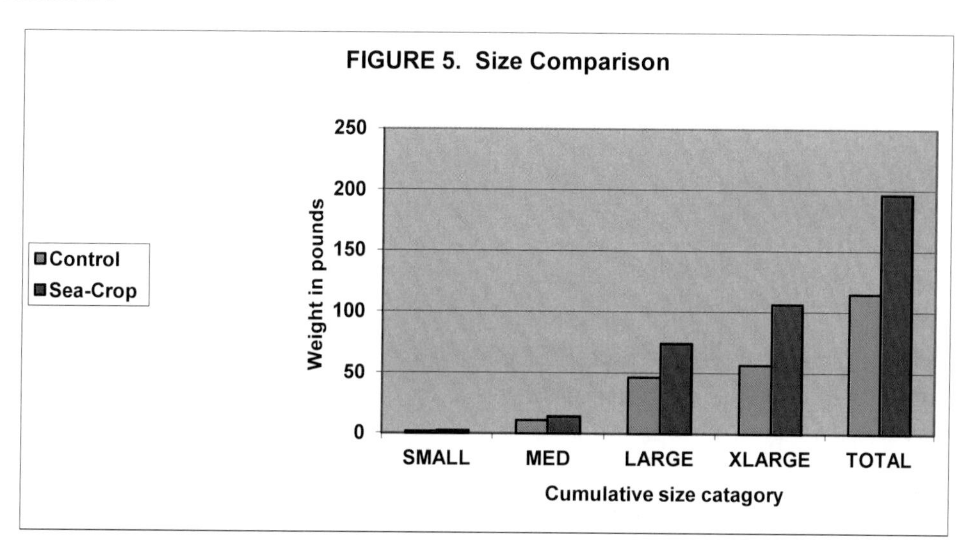

FIGURE 5. Cumulative Size Category (MED=medium, XLRGE= extra large) Weights (in pounds) of Tomato Fruit across 6 Harvests for 2008 Cal-Agri Trial as Influenced by the treatment.

The bar graph that follows shows how root growth was improved by a SEA-CROP® treatment of zucchini grown in coco shell soilless medium, in a commercial greenhouse, in Morocco by Dr. Abdelhaq Hanafi.

Impact of SEA-CROP® on Root of Zucchini cv. Apus grown in Soilless system

Dry Weight (g) of 8 Root Systems/Treatment

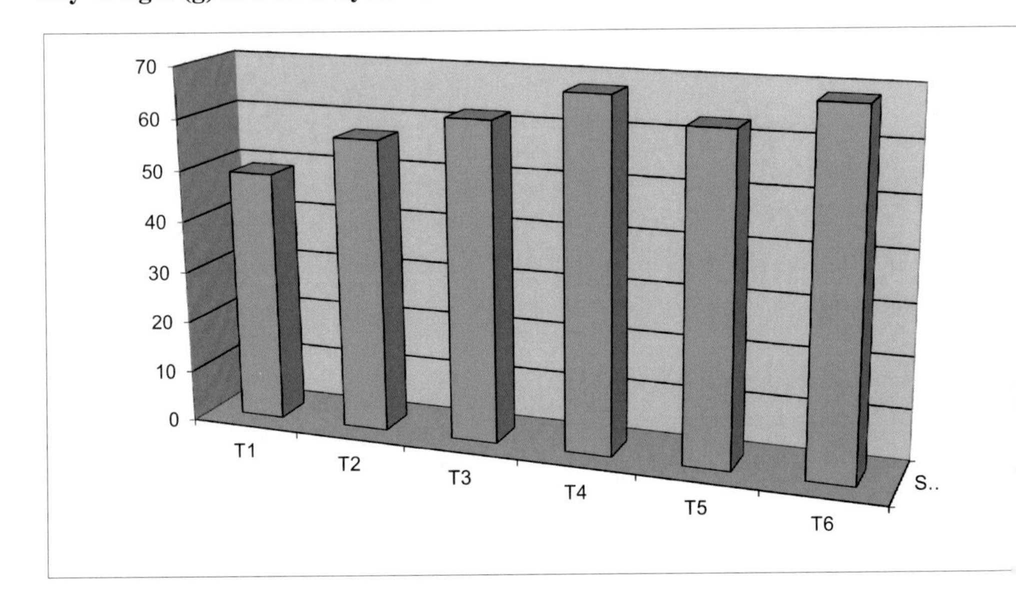

Treatment

T1: Negative control

T2: SEA-CROP® applied once as Drench (0.15%)

T3: SEA-CROP® applied once as Drench (0.30%)

T4: SEA-CROP® applied twice as Drench (0.30%) (6 weeks apart)

T5: SEA-CROP® applied once as Foliar (1%)

T6: SEA-CROP® applied twice as Foliar (1%) (6 weeks apart)

This photograph shows how root growth was improved by a SEA-CROP® treatment of green beans grown in a commercial greenhouse in Morocco by Dr. Abdelhaq Hanafi.

The left hand holds the root system of an untreated control. The right hand holds a plant that received two treatments of 1% strength SEA-CROP®, one soil drench treatment and one foliar application.

From the same green bean study in Morocco at Agidar, this photo shows that there is a dose/ benefit relationship when treating with seawater concentrate.

The row of plants on the left received two treatments of 1% strength SEA-CROP®, one soil drench treatment and one foliar application. The row on the right received only a single soil drench treatment containing the very low concentration of 0.15% SEA-CROP®.

Confirming the above result, a dose/ benefit study done on cucumbers by Dr. David Nielson at Davis, California showed that additional benefit was achieved with each additional increment applied up to 10 gallons of SEA-CROP® per acre.

The photo below shows test results obtained by Cal-Agri in Eugene, Oregon.

Rosemary (Salem) starts were transplanted into one gallon pots and treated with SEA-CROP® at the rate of 3 gallons per acre. Cotton seed meal was used on both the untreated control and treated tests as an organic fertilizer. After 30 days the plants were dug up, the roots washed and the entire plant desiccated so that dry matter weights could be compared. The percentage gain of the SEA-CROP® treated plants as compared to the controls is listed below.

Start Date	Duration	Replications	Dry Matter Increase of SEA-CROP® Treated Plants		
			Roots	Tops	Total
7-19-07	45 days	6	23.14%	15.29%	17.24%

From a personal trial by the agricultural consultant Lawrence Mayhew

"The photos of the strawberries are an example of what happened. The first photo shows some strawberries approaching the ripening stage."

"The second photo shows the treated strawberries, which doesn't seem too exciting at first glance unless you realize that not only are the SEA-CROP® treated berries ripening sooner than the untreated, they are the second crop, while the unripe berries are only yielding their first crop."

Soy Beans are inspected on a farm in Malawi, Africa. Notice the fuller root development and greater number of pods evident on the treated cluster to the left as compared to the untreated cluster on the right. The treated and untreated test plots were side by side in the same field. Each cluster represents one planting station.

Groundnuts (peanuts) are inspected on a farm in Malawi, Africa. Notice the fuller root development and greater number of pods evident on the treated cluster to the left as compared to the untreated cluster on the right. The treated and untreated test plots were side by side in the same field. Each cluster represents one planting station.

Another photograph from the peanut trial in Malawi, Africa. The cluster of groundnut (peanut) plants in my left hand were from a test plot treated with four gallons per acre of SEA-CROP® seawater concentrate and the ones in the right hand were from an untreated control plot.

Overall, extra podding was observed at the rate of 20%. Final increased yield for the crop at harvest was measured at 42.8% greater than the untreated control. It should be noted that the soils at this estate are quite acid, having a pH of 4.51.

Chitala Research Station is part of Chitedze Agricultural Research Station of the Department of Agricultural Research Services (DARS) which is within the Ministry of Agriculture & Food Security of the Government of Malawi.

At the Chitala Research Station the treated corn on the left has good kernel development all the way to the top of the cob. The kernels are larger and there are more rows of kernels than the normally fertilized control on the right. SEA-CROP® enables plants to more nearly achieve their full genetic potential as demonstrated in this example.

Leaves and raspberries from an untreated raspberry plant on left and SEA-CROP® plant on right. The 6 berries from the treated plant weighed 20.66 gm and the 6 berries from the control weighed 8.53 gm.

While these leaves and berries represent the same variety from the same nursery and planted on the same day, the larger ones are from a plant that had been treated with SEA-CROP® for three years.

When treating perennials, especially mature trees, it may be several years before maximum results are obtained.

These apples were brought to the author by a third party. They are from a young orchard and were said to be representative of both the treated and untreated portions of that orchard. These trees had not been treated with seawater concentrate in previous years.

Apple, pear and citrus orchards treated with SEA-CROP® have experienced healthier trees and reduced cull rates.

These treated strawberries were grown during the Cal-Agri Products field trials at Louisiana State University at Baton Rouge in 2008. In this trial, SEA-CROP® was used as a root dip on plug transplants at the dilution rate of one part in fifty (2%). That single treatment gave a 25% increase in yield compared to the untreated control.

Seed date- 10/19/2011

Soil drench application date - 10/20/2011, 2 gallons per acre SEA-CROP® @ 5% strength co-applied with 2 gallons per acre molasses @ 5% strength.

Emergence - 10/26/2011

1st foliar app date -11/16/2011, 1 gallon per acre SEA-CROP® @ 2% strength.

2nd foliar app date - 12/7/2011, 1 gallon per acre SEA-CROP® @ 2% strength.

Harvest date - 2/13/2012, **Duration of growing cycle** -118 days

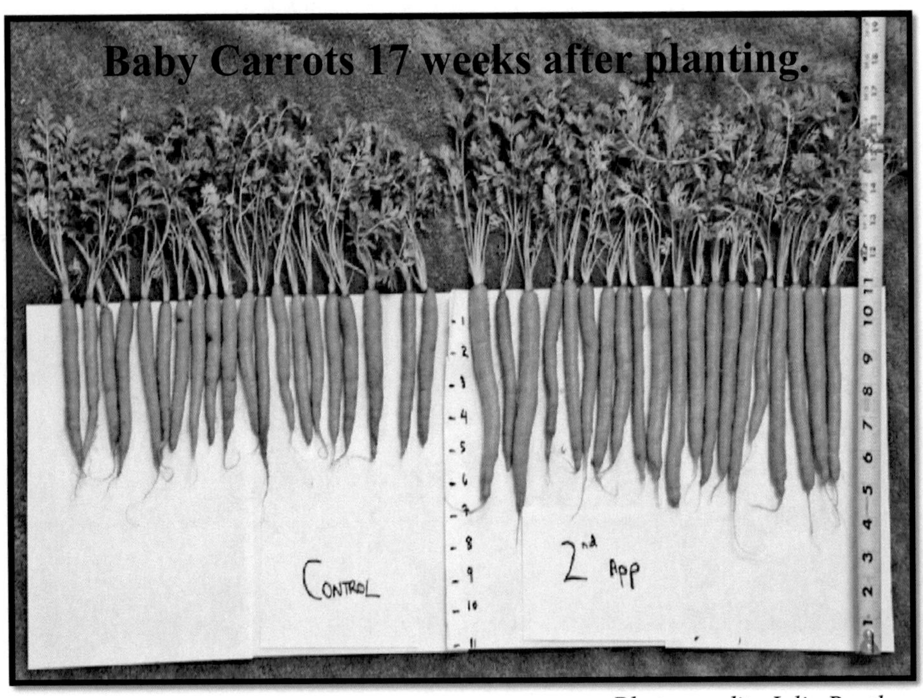

Photo credit- Julie Brothers

Marketable Yield % increase	**41.57%**
Brix Refractometer reading increase	**30%**
Insect Damage % decrease	**33%**

Baby Spinach Harvest Results March 2012 @ 57 days

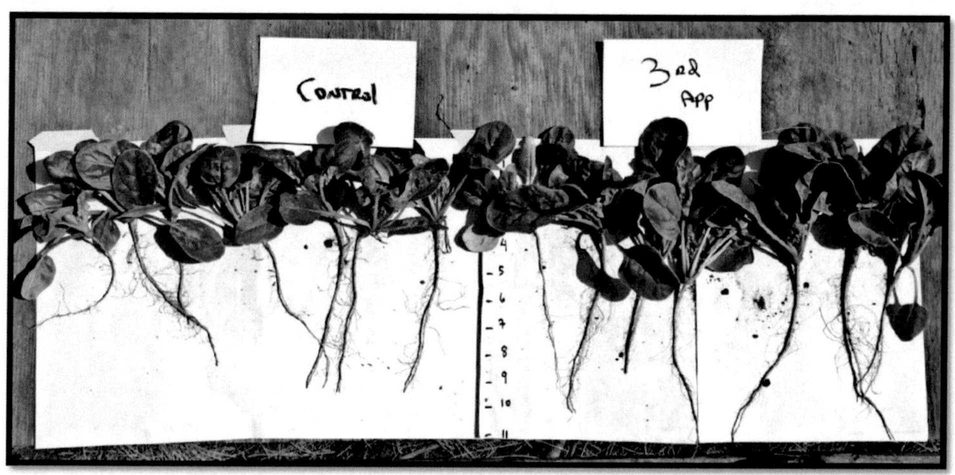

Photo credit- Julie Brothers

% Increase Over Control

Catagory	**2 applications** SEA-CROP®	**3 applications** SEA-CROP®
Brix *(average of 4 plots)*	+28.6%	+23.8%
Fresh Weight *(greens portion) 35 representative plants of each category*	+43%	+70.1%
Dry Matter Content Increase	+11.8%	+ 5.9%

If a picture is worth a thousand words then the following two photographs should be worth two thousand. The side by side photos show winter wheat plantings in Germany. In the field on the left glyphosate has only been used for one year and has had no opportunity to build up in the soil. In the field on the right glyphosate has been used for ten consecutive years.

Dr. Bott, U. of Hohenheim, Stuttgart, Germany

Field observations in winter wheat production systems made in Southwest-Germany in 2008 & 2009 point to potential negative side-effects of long-term glyphosate use.

Chapter Nine

How Plants Shouldn't Grow

"All pesticides and herbicides, even the natural ones, have undefined or unintended consequences." ~ Doctor Dwayne Beck

"The poison was brewed in these West lands but it has spat itself everywhere by now." ~ C.S. Lewis

Glyphosate, also known as Roundup® is a witch's potion brewed in the cauldrons of Monsanto Chemical Corporation. Born in the USA, it is now used extensively around the world and is even being imposed upon third world countries, on the brink of starvation, as a solution to their food security problems.

For example, it is being promoted for use in Malawi and other African nations by the Gates Foundation. The Gates Foundation spent $27.6 million on 500,000 shares of Monsanto stock between April and June 2010.[1] This might have something to do with the reason that they are now promoting the use of genetically modified Roundup® Ready corn and cotton plantings throughout Africa.

In Malawi, 80% of the population are subsistence farmers on family plots of land that average about two acres in size. Riddled with malaria, HIV/AIDS and malnutrition, Malawians are one bad harvest away from death.

Every 30 minutes in India a farmer commits suicide and a favorite method is by drinking Roundup®. Many of these farmers became impoverished by first buying expensive, Monsanto produced, Roundup® Ready GMO cotton seed and then sustaining crop failures when the seed failed to perform as advertised especially under drought conditions. Suicide by Roundup® is the only method of protest left to these poor souls who feel that their lives have been destroyed by Monsanto Chemical Corporation.[2]

Glyphosate compromises and poisons a plants immune system through the instrumentality of chelation which ultimately leads to the plants death.

Chelation - The word chelation is derived from Greek word *chele*, meaning "claw"; a chelating agent surrounds the central metal atom as if it were caught in the claws of a lobster.

It involves chemicals that form soluble, complex molecules with certain metal ions, inactivating the ions so that they cannot normally react with other elements or ions.

A metal atom that has been caught by a strong chelating agent is not available to a plant for use as a metallic cofactor in enzyme formation.

Glyphosate was first patented in 1964 as a chelating agent. It is a very strong chelating agent and that is how it poisons plants. Monsanto filed a patent in1974 for use of glyphosate as a weed killer. When plants are poisoned with Roundup ®, normal enzyme activity cannot take place and ultimately some enzymes cannot even be formed. The plant's immune system becomes completely compromised to the extent that in a very short time it is invaded and destroyed by bacterial pathogens living in soil in which the plant grows.[3]

Glyphosate makes plants more vulnerable to disease. It has been known since the 1980s that Roundup® destroys plants by knocking out their immune system, allowing soil bacteria to complete the destruction. Seedlings grown in a sterilized medium, such as sterilized soil or in sterilized soilless substrate, are not killed by normal glyphosate applications. It takes 45 times the normal application rates to kill plants raised in this manner. That clearly shows how the poison works. It prevents plants from uptaking and utilizing the irreplaceable trace minerals that they need for making the enzymes necessary for health and growth.

Plants are not the only things that glyphosate kills. It poisons beneficial insects that kill other species that are agricultural pests. The International Organization for Biological Control found that exposure to freshly dried Roundup® killed over 50% of three species of beneficial insects: parasitic wasps, lacewings, and ladybugs. It has also been demonstrated that Roundup annihilates beneficial soil organisms such as earthworms and nitrogen-fixing bacteria.[4]

It has been shown that repeated applications of glyphosate significantly affect growth and survival of earthworms. Biweekly applications of low rates of glyphosate (1/20 of typical rates) caused a reduction in growth, an increase in the time to maturity, and an increase in mortality.

Glyphosate and glyphosate-containing products have also been proven to kill a variety of arthropods. Over 50 % of some beneficial species were killed by exposure to Roundup®. These small creatures are important in humus production and soil aeration.

As detailed in Chapter 4, it is important to remember the role of microflora symbiosis in its rhizosphere to a plant's proper function. About a fourth of the glyphosate adsorbed by the plant is exuded from the root system. In the presence of glyphosate, a plant is forced to exude poison.

Glyphosate does not remain on the surface of a plant on which it has been sprayed. It penetrates and permeates every part of the plant including the all-important root exudates. The poison released, in what should be nourishing root secretions, is directly toxic to whole classes of beneficial bacteria and fungi, including mycorrhizae. There are ten tons of living materials in an acre of soil. Soil is a living entity and the quality of the life it contains determines what minerals are available for uptake as nutrients. Glyphosate takes entire groups of organisms out of the soil.[5]

In recent years, Monsanto has been promoting the use of glyphosate for an insidious form of no-till agriculture. Farmers are urged to "burn down" weeds and cover crops with herbicide rather than turning them under by plowing. This practice has the potential for disastrous consequences.

For decades Monsanto insisted, and government agencies agreed, that glyphosate was inert and completely safe once it hit the soil, that it had no residual activity. Supposedly it became so tightly bound to clay particles that none could be taken up by a crop planted in a freshly "burned down" field. In addition, Monsanto offered "proof" that Roundup® had a half-life in the soil of from 3 to 60 days. They claimed that even the residual that was tightly bound up in soil was rapidly degraded, so that the poison was actually harmless and environmentally friendly. Unfortunately folks it just ain't so.

It is not unusual when examining agricultural soil under a microscope to see undecomposed crop residue from several previous plantings. Roundup® permeates every part of a plant. As the plant decomposes in soil glyphosate is released in plant-available form to poison the various life forms that it may contact.

In addition, glyphostae bound to clay or humus particles was said to be completely inert until it was harmlessly decomposed and then disappeared in short order without a trace. Testing by a number of scientists has shown that to be untrue on a number of counts.

First of all, testing in countries outside of the United States and beyond the influence of the eight million dollars that Monsanto spends directly on lobbying each year has shown that glyphosate has a much longer half-life than stated by Monsanto. In fact, it can be measured in soils as long as three years after a single application. Degradation of glyphosate in most soils is slow, to practically nonexistent. Repeated applications over a period of years can cause a buildup of the broadly acting poison with very serious and long-lasting effects.[6]

Secondly, the reputedly "deactivated" glyphosate adsorbed onto clay and humus particles has been shown to desorb and be released in plant-available form when phosphorus fertilizers are applied to the soil. Phosphate likes the same sorption sites on soil particles that glyphosate does. It likes them better so it displaces the poison, which can then take as long as two weeks to find a new home on other available sorption sites. In the meantime it is loose and available to destroy the immune system of any plant with which it comes in contact.

The third way in which reality differs from the imagined or hoped for safety of Roundup® has to do with its decomposition. Not only does it take much longer to decompose than advertised, one of its major decomposition products is a chemical with the jaw-breaking name, aminomethylphosphonic acid which, for good reason, is referred to as **AMPA**. It turns out that AMPA, just like its parent glyphosate is very resistant to decomposition and persists for long periods in the soil. It is also plant toxic. In some ways AMPA is even more toxic than glyphosate because it affects Roundup® Ready plants as well as plants with unmanipulated natural genetics. AMPA has been proven to reduce the germination and vigor of seeds planted in contaminated soil.[7]

According to Monsanto Chemical Corporation and the USDA, Glyphosate has a half life of six weeks. Do you believe it? Do you think that the half billion dollars that Monsanto has given in political campaign contributions in the last decade might have had an influence on the USDA?

False Advertising: On Fri Jan 20, 2007, Monsanto was convicted of false advertising of Roundup® for presenting it as biodegradable, and claiming it left the soil clean after use.[8]

Scientific Fraud: On more than one occasion, the United States EPA has caught scientists deliberately falsifying test results at research laboratories hired by Monsanto to study glyphosate.[9]

Amino Acids, Peptides, and Proteins

Amino acids are the basic building blocks of enzymes, hormones, proteins and body tissues in both plants and animals. They are basically carbohydrate molecules that have nitrogen as a component. A peptide is a compound consisting of 2 or more amino acids. Peptides consisting of more than 50 amino acids are classified as proteins. The distinction between peptides and proteins is arbitrary and based on size. Enzymes, as we learned in Chapter 3, are the special proteins that are required for every biological operation necessary for life to exist. Most enzymes require an atom of metal in their core in order to perform their special biological function.

Roundup ®works its magic by inhibiting an enzyme involved in the formation of amino acids. If amino acids cannot be formed then of course peptides, proteins, enzymes and hormones cannot be synthesized either.

In the year 2000, Monsanto's US patent on glyphosate expired and that allowed competition by other companies wishing to manufacture and market the chemical. As a result, prices dropped dramatically. The plant killer became relatively inexpensive to use and of course less profitable for Monsanto but Monsanto Chemical Corporation had a plan.

In 1996 Monsanto released "Roundup® Ready" (RR) soybean seed into the marketplace. It had been found that certain bacteria produced an enzyme that was resistant to the effects of glyphosate and would prevent the poison from initiating the cascade of destructive chemical process that normally occur in treated plants.

DNA from the bacteria was genetically engineered into the Roundup® Ready (RR) soybean seeds.

RR seeds for corn, canola, cotton, sugar beets, wheat and alfalfa soon followed. These seeds all have inserted genetic material from viruses and bacteria that allows the crops to withstand applications of normally deadly Roundup®.

Monsanto requires farmers who buy Roundup® Ready seeds to only use the company's Roundup® brand of glyphosate. This has extended the company's grip on the glyphosate market, even after its patent expired in 2000. Farmers in the USA now drench more than 135 million acres of Roundup® Ready crops with Roundup® each year and the rate of use is accelerating annually. Unfortunately the rapid increase in glyphosate use is not limited to the United States. It is a worldwide phenomenon.

China has recently become the leading producer of glyphosate in the world and as of 2010 now has the capacity to produce 835,900 tons per year. At the recommended rate of 22 ounces per acre, the production of China alone can now produce enough to treat one billion three hundred thirty-seven million (1,337,000,000) acres of the world's prime agricultural cropland.

As instrumentation and analytical methods have improved, as suppressed research has come to light, and as decades of planting experience has broadened the base of real world results, it has begun to appear that much of the best croplands in the world are being subjected to degradation. Even if glyphosate use stopped immediately it would take decades to reverse this poisoning.

Glyphosate persists in soil for much longer periods of time than Monsanto has been willing to acknowledge. The concentration in the soil builds up season after season with each subsequent application. Their own test data revealed that only 2% of the product broke down after 28 days. Whether it degrades in weeks, months, or years varies widely due to factors in the soil, including pH, temperature, clay, types of minerals and the amount of residues from Roundup® Ready crops. In some conditions, glyphosate can grab hold of soil nutrients and remain stable for long periods of time. One study showed that it took up to 22 years for glyphosate to degrade only half its volume from a single application. Needless to say, yearly applications under such conditions would cause a buildup in the soil of drastic proportions and poison the soil for decades to come so that only RR crops would grow in it and those not very well.[10]

Glyphosate builds up year after year in the tissue of perennial crops like alfalfa that get sprayed repeatedly. In 2011, the USDA approved Monsanto's gene-modified RR alfalfa for planting in the USA. Alfalfa is the fourth largest crop in the US, grown on 22 million acres. It is used primarily as a high protein source to feed dairy cattle and other ruminant animals. Even without the application of glyphosate, the nutritional quality of Roundup® Ready alfalfa will be less, since all Roundup® Ready crops, by their nature, have reduced mineral and other nutritional content.

Roundup®, by chelating and sequestering trace minerals, affects more than just one plant enzyme and the genetic material inserted into RR seeds only conveys immunity to a portion of the consequences of using Monsanto's magic potion. The following list shows the reduction of mineral content in three year old Roundup® Ready alfalfa, after it had a single application of glyphosate the previous year. The comparison is made to non-RR alfalfa.[11]

% Reduction of Alfalfa Nutrients by Glyphosate	
Nitrogen	13%
Phosphorus	15%
Potassium	46%
Calcium	17%
Magnesium	26%
Sulfur	52%
Boron	18%
Copper	20%
Iron	49%
Manganese	31%
Zinc	18%

Roundup® Ready crops dominate US livestock feed. Soy and corn are most prevalent as 93% of US soy and nearly 70% of corn is Roundup® Ready. Animals are also fed processing derivatives of three other Roundup® Ready crops: canola, sugar beets, and cottonseed. Nutrient loss from glyphosate can therefore be severe.

There are veterinarians who believe that the health of America's farm animals has been greatly impaired since RR crops have come to dominate the available feedstock. Unfortunately, no one is tracking this, nor is anyone looking at the impacts of consuming milk and meat from GMO-fed animals. One thing that we do know is that the RR crops being produced on America's glyphosate poisoned farmland are in a rapidly escalating state of poor health.

Dr. Don Huber spent 35 years as a plant pathologist at Purdue University and knows more than a little bit about what makes plants sick. He is also one of the world's experts on the negative impact of Roundup® on crop species, both RR and conventional. By weakening plants and promoting disease, glyphosate opens the door for lots of problems. According to Doctor Huber, *"There are more than 40 diseases of crop plants that are reported to increase with the use of glyphosate, and that number keeps growing as people recognize the association between glyphosate and disease."*

Besides weakening plant defenses and increasing pathogen populations and virulence, glyphosate can have indirect effects on predisposition to diseases resulting from immobilization of micronutrients involved in disease resistance, reduced growth and vigor of the plant and modification of the soil microflora affecting the availability of nutrients. It appears that as Roundup® use increases plant disease skyrockets.[12]

It is well established that spraying glyphosate promotes plant diseases. In addition it may also promote insect infestation because many bugs are opportunistic and seek sick plants with weak immune systems. It is known that healthy plants produce nutrients in a form that many insects cannot assimilate. Farmers around the world report less insect problems among high quality, nutrient-dense crops. Weaker plants, on the other hand, are a bug feast just waiting to happen. Plants ravaged with diseases promoted by glyphosate may also attract more insects, and that in turn, will increase the use of toxic pesticides that further destroy beneficial links in the soil-food-web.

There is a possibility that eventually so much glyphosate will accumulate in soils that arable land will become too toxic to grow anything. It is important to remember that there was such a thing as successful farming before Monsanto, before RR seeds and before the universal use of herbicides.

In addition to lower nutrient content, do Monsanto's Roundup® and Roundup® Ready GMO food crops pose other health hazards to humans as well as animals? It appears that they may do so.

The latest damaging evidence comes from the United States Geological Survey (USGS), which is now detecting glyphosate in streams, the air we breathe and even rain. While the concentrations detected in rain and air are small, emerging scientific evidence about what these chronic low-level exposures do to our bodies, particularly among unborn babies and young children, is cause for major concern.[13]

Pesticide-exposure expert Warren Porter, PhD, professor of environmental toxicity and zoology at the University of Wisconsin, Madison took the air exposure numbers from the USGS study and found reason for concern. His calculations showed that the levels found in the USGS survey could lead to accumulations that might alter endocrine mediated biochemical pathways, leading to obesity, heart problems, circulation problems, and diabetes. Low-level exposure to hormone disruptors like glyphosate has also been linked to weakened immune function and learning disabilities.

Why is glyphosate now in the air we breathe? The majority of corn, cotton, canola, and soy crops grown in the United States are genetically engineered to tolerate spraying with Roundup®. So much of this chemical is now being used that it is beginning to permeate our environment. When we eat those crops or when they're turned into ingredients used in processed foods, we wind up eating the Roundup® too. Not only are we breathing it in and getting soaked in it when it rains, but we're also eating it at every meal and with every snack.

Glyphosate has been associated with deformities in a host of laboratory animals and it has been reported that Monsanto has known about Roundup®'s link to birth defects since the 1980s, when internal research found mutations in animals exposed to high doses. It has also been linked to sterility, hormone disruption, abnormal and lower sperm counts, miscarriages, placental cell death, birth defects and cancer.[14]

The same nutrients that glyphosate chelates and prevents plants from utilizing are vital for human and animal health. These include iron, zinc, copper, manganese, magnesium, calcium, boron, and others. Deficiencies of these elements in our diets are known to interfere with vital enzyme systems and cause a long list of disorders and diseases. Alzheimer's, for example, is linked with reduced copper and magnesium and the incidence of this tragic disease has increased 9,000% since 1979.

Glyphosate, even in plants genetically engineered (GE) to withstand it, affects about 25 different enzymes in the process of chelating critical micronutrients that make the plant function properly. According to Dr. Huber, the nutritional efficiency of genetically engineered plants is profoundly compromised. Micronutrients such as iron, manganese and zinc can be reduced by as much as 80-90% in GE plants.

From this lack of mineral nutrient density in GE plants there is a malnutrition ripple effect that carries over to soil microbes, animals and humans. This does not even factor in the potential effects of having glyphosate in the food chain.

Finally, glyphosate interferes with the beneficial bioactivity of seawater concentrate. It chelates metallic ions crucial to the plants ability to form the enzymes and prosthetic groups it needs to enable it to more nearly achieve its full genetic potential. Nonetheless, a way has been found to overcome the problem and in many cases give valuable assistance to the farmer who is enmeshed in Monsanto's web. The following bulletin was issued to all of Ambrosia Technologies' SEA-CROP® distributors in 2010:

CORRECTING FOR GLYPHOSATE 11/22/2010

Research recently published by Dr. Don Huber, professor emeritus, of Purdue University, must be taken into account when SEA-CROP® will be used in a field on which Roundup® or other glyphosate treatment will be used or has been used within the last 18 months.

Dr. Huber's findings, from twenty years of investigating glyphosate, show the following results that could negatively impact SEA-CROP®'s bioactivity.

1. Glyphosate is a very strong chelating agent that sequesters minerals and can tie them up indefinitely.

   Remedy: Glyphosate should not be used within two weeks before or after the application of SEA-CROP®.

2. When glyphosate is used for "burning down" weeds before planting, free glyphosate may be released in phytotoxic amounts from the decomposing rootlets of the "burnt down" weeds.

   Remedy: SEA-CROP® should not be used within two weeks of "burning" down with glyphosate.

3. Free glyphosate binds to organic matter in soil and persists for a long period of time. When phosphate is applied to the soil, as in NPK fertilizer or other forms of phosphate, glyphosate is displaced by the phosphate and once again becomes free and phytotoxic.

   Remedy: SEA-CROP® should not be used closer than two weeks before or after the application of NPK or other phosphate fertilizers.

If glyphosate was used to "burn down" weeds or cover crops just prior to planting then it is best to wait until after emergence to treat the crop.

Diluted SEA-CROP® may be sprayed over the row on soybeans when four to five inches high and on corn when six to eight inches high. Set the width of the spray at twelve to eighteen inches wide and dilute the SEA-CROP® to a concentration no greater than ten percent.

Molasses may be co-applied with the SEA-CROP® at the rate of two gallons per acre. Other sources of sugar, such as dextrose, lactose or sucrose, may be substituted for the molasses at the rate of six to ten pounds per acre or, alternatively, unpasteurized skimmed milk may be used instead of sugar at the rate of one to six gallons per acre.

Farmers using these procedures have been able to decrease nitrogen inputs and significantly increase yields despite the presence of glyphosate.

In closing this chapter, it should be repeated that there is a growing body of evidence that glyphosate builds up in soils and under some conditions may have a half-life of up to three years or more. This means that the residual buildup from repeated annual applications is considerable.

The photograph of winter wheat on page 86 of this book is a graphic illustration of the consequences.

Chapter Ten

The Future of Seawater Concentrates

"When it comes to the future, there are three kinds of people: those who let it happen, those who make it happen, and those who wonder what happened."

~ *John M. Richardson, Jr.* American academic

In chapter seven, "How Plants Shouldn't Grow", we saw an apocalyptical vision of the future, a future brought to us by Monsanto Chemical Corporation and their ilk. In Monsanto's vision of the future all seed will be genetically modified so as to be able to grow in ground that will become increasingly poisoned by glyphosate and, of course, the GMO seeds and poison must be obtained from Monsanto.

Once a farmer has started to plant Roundup Ready seed and apply regular applications of glyphosate, he is in a condition similar to that of a junkie who has become addicted to shooting up heroine. Each time the action is repeated it can only lead to further destruction of all that is best in his life. Yet it is difficult to quit.

His precious soil cannot sustain such terrible abuse and remain fertile. In time the farmer's soil will become no more than a sterile substrate for roots to clutch onto so that plants can remain upright and the farmer will be engaged in nothing more than a form of outdoor hydroponics.

Crops produced in such a manner are not fit as food for man or beast. No man should eat such food nor should he raise it to feed to his neighbor and fellow man. Such crops may be useful for energy production by burning them or for alcohol production but, they are no longer food crops.

In such a future of farming, bacteria will not be fixing nitrogen from the air nor will fungi be leaching phosphorus from rock minerals for plants to use. All nutrients will have to be mined in remote locations and transported to the farmer's outdoor hydroponic operation. The worn-out, poisoned, sterile soils will not be capable of giving nutrients to the plants grown in them. All nitrogen inputs will have to come from petroleum feedstock transported from half way around the planet.

Worse yet, "drag" with GMO crops is a well-known phenomenon. GMO plants are not healthy plants. According to Dr. Don Huber they are subject to increased disease conditions from as many as 40 different disease pathogens that their natural plant relatives take in stride.

This is a nightmare future that cannot be long sustained nor can such a system of agriculture sustain the current population of planet Earth. Our prime agricultural lands worldwide are in peril from such a system of agriculture and the inevitable result must eventually be worldwide famine.

Conventional farming publications are filled with visions of a new, high-tech future for farming. Technology is seen as the key. Genetically modified Roundup Ready crops of all kinds are hailed as labor-saving solutions to the future of farming. The vision is false. The cost of inputs alone makes this scenario impossible. As soil becomes poisoned, worn out crops will need to be spoon-fed scarce nutrients brought in from mines scattered around the world. Competition for these mineral nutrients will become fierce and prices will necessarily skyrocket. Historically competition for scarce resources has been a prime cause of warfare.

This need not be the future. There are alternatives available that can be chosen now.

A Better Future.

There is another path that can be taken. Damaged soils can be brought back to life and nutrient dense food can be made available to all. Living soils teeming with microbes and fungi can, with only a little help, provide plants with all the nutrients they need.

One of the keys to a vital and abundant agricultural future is seawater concentrate. Seawater covers over seventy percent of the surface of this planet. It is the least scarce resource that we have. It is virtually inexhaustible because its mineral content is constantly replenished as the rivers of the world dump billions of tons of minerals into the oceans each year. Undersea volcanoes, hot springs and volcanic eruptions on land also add billions of tons of elements. Ninety percent of all volcanic activity on Earth occurs in the oceans.

Instead of assassinating soil biota with poison and tortured genetics, they can be encouraged to be energetic symbiotic partners. In a healthy soil-food-web the microflora do all of the heavy lifting.

Seawater concentrates applied in practically homeopathic amounts have been documented to build up soil microflora populations. They have also been proven, when used properly and consistently, to enable high yields of nutrient dense crops.

A farmer has no choice other than to follow the dictates of economic reality. That also means that if he is to have a future in farming he must preserve his capital and the most precious material asset that he possesses is his soil.

Seawater concentrates have been proven to improve soil tilth in just a few short years, in addition to providing consistently high yields in a sustainable manner. Again, it is the plants and the soil biology that do all of the heavy lifting. Seawater concentrate acts as a catalytic trigger that enables both plants and soil biology to more nearly achieve their full genetic potential.

Slash-and-burn agriculture is now considered to be an artifact of the most primitive cultures. Yet in its day, it was an advanced solution for exhausted and worn out soils. Burning off vegetative cover released mineral nutrients for availability to crops planted in the newly cleared land. As available nutrients were depleted, good yields of nutritious crops could be maintained by simply abandoning fields and clearing new ones, burning off the slash and planting anew in an endless cycle.

Crop rotation was unnecessary and endless, repeated plantings of the same crops, such as corn or tobacco, took place each planting season. With limited population, with unlimited forest and jungle for clearing, this was a system that worked very well. Global populations today are much larger and fertile agricultural lands are no longer available just for the labor of clearing and burning. Loading up the covered wagon and heading West to find virgin lands to clear is no longer an option.

As a culture we are only a few generations removed from slash-and –burn agriculture. Moving onward frequently to more fertile land is no longer a way of life or even possible. The farmer now remains in one place and if he wishes to farm fertile land he must enable the soil to regain and maintain fertility.

This is the situation that has given rise to chemical agriculture as it exists today. The concept of replacing the minerals extracted and sold off the farm each year is not necessarily a bad one. The problems arise from the very incomplete understanding we possess of the great complexity and variety of the soil-food-web and the beneficial symbiosis between plants and soil biota.

It may be that Monsanto Chemical Corporation and their ilk are not truly as evil as they appear to be. As Upton Sinclair once said: "It is difficult to get a man to understand something when his job depends on not understanding it." It may be that they are living in a state of denial and willful ignorance but that is no excuse for the rest of us.

The organic soil component contains all the living creatures in the soil plus the dead ones and plant debris in various stages of decomposition. An acre of living soil can contain 900 pounds of earthworms, 2400 pounds of fungi, 1500 pounds of bacteria, 133 pounds of protozoa, 890 pounds of arthropods and algae, and even small mammals. In fact, soil should be viewed as a living entity rather than an inert medium for roots to clutch so that plants stay upright.

Crops raised and then sold off the farm each year may contain as many as 60 chemical elements in their makeup and these need to be replaced, just the same as nitrogen, phosphorus and potassium. Neither plants raised in the soil, nor the soil biota that dwells therein, can be healthy and thriving in an environment that is mineral depleted. Crop residues that feed soil microflora need to supply complete nutrition or nature's system of soil fertility starts to falter at the fundamental level.

The idea that only three minerals need to be replaced or even a dozen minerals is flawed and old fashioned. It is an outdated concept based on ignorance. A Mozart Concerto can't be played using only three notes of the musical scale and nutritious food crops can't be grown with three chemical elements when 89 elements exist in nature.

Appropriate mineral nutrition needs to be present for soil organisms and plants to prosper. Adequate levels of all minerals that exist in nature should be supplied and the most convenient and economical method is to supply them in the form of seawater concentrate.

During the last century man's understanding has advanced on many fronts and, hindsight being better than foresight, from our present position in time it is easy to recognize some of the mistakes that our predecessors may have made. No doubt posterity will consider us to have been dwellers in semi-ignorance as well.

Nearly a century of experimentation with seawater has shown that it is a wonderful resource that can and must be utilized for more than just making sea salt. Modern methods of making seawater concentrate and years of agricultural trials learning how best to use those concentrates in agriculture, have now made possible an inexpensive and effective means for farmers to enrich and protect their soil from depletion and exhaustion.

The future of successful agriculture must, of necessity, involve remineralization of agricultural soils and the least scarce resource to supply those minerals are the seas of the Earth. Seawater concentrates have the potential to offer relief and a new way of farming to those farmers who want to get off of Monsanto's treadmill.

Farming with seawater concentrate is the modern way to supply complete mineral nutrition to both crop plants and soil biota. Seawater concentrate is the key to a future of agricultural abundance.

Afterword

Sing to the Lord of harvest, Sing songs of love and praise;

With joyful hearts and voices Your alleluias raise.

By him the rolling seasons In fruitful order move;

Sing to the Lord of harvest, A joyous song of love. ~ John S B Monsell

The task is over, the book is written and, for better or worse, it is now out into the world in its present form. I hope that the reader will forgive any defects and appreciate the value of its content and the message of seawater concentrate.

Sir Isaac Newton once said: "To explain all nature is too difficult a task for any one man or even for any one age." That is certainly true and in this small book the author has labored to convey his growing understanding of a very complex subject.

In a book of this size and scope, of necessity some subjects have been simplified and others glossed over. Nonetheless, I hope that the reader has gained increased knowledge and understanding of current conditions in agriculture and the need for a new path to be taken.

The potential for seawater concentrate is well documented and the information presented in this book represents only a very small portion of the work that has been done by researchers in this field.

Hopefully the readers' curiosities have been stimulated to the extent that they will try seawater concentrate and experience for themselves how good food can taste when it has been raised in the presence of all the elements that God put into his good earth.

APPENDIX A

Case Studies

Dose/Benefit Study

A dose/ benefit study done on cucumbers by Dr. David Nielson at Davis, California showed that additional benefit was achieved with each additional increment of product applied up to 10 gallons of SEA-CROP® per acre.

In the Abstract of the study Dr. Nielson stated:

Sea-Crop® is a seawater extract product that is used as a soil additive or foliar treatment to promote plant growth. Here we examine the safety and efficacy of this product when applied as a soil drench across a wide range of applications to potted cucurbits. All treatment rates, excluding the highest concentration, showed excellent plant safety. Most product rates enhanced both foliar and root growth, increasing average dry weight by more than 50% in many cases. This trial clearly shows promising plant growth and safety results for soil application of Sea-Crop to young cucumber plants.

County Line Farm 2012 Winter Trials by Collé Agriculture

This study demonstrated that when six different species of vegetables were treated with a split application of SEA-CROP® at the rate of 4 gallons per acre the following average results were obtained:

Yield increase	+62.56%
Dry matter increase	+12.72%
Brix increase	+35.12%
Protein increase	+19.73%
Insect damage reduction	52.%

The data summary for each species follows.

County Line Farm Beet Harvest Results April 2012 (report summary)

Drench date - 10/11/2011 Seed date- 10/10/2011
Emergence- 10/2 /2011
1st app date - 12/08/2011
2nd app date - 12/29 /2011
3rd app date- 1/19/2012
harvest date – 4/03/2012
Duration of growing cycle – 176 days

Brix

Refractometer readings

Portion	Control	2 applications	3 applications
Greens	6	10	9
% increase over control		*+66.7%*	*+50.0%*

Portion	Control	2 applications	3 applications
Root	8.0	11.0	14.0
% increase over control		*+37.5%*	*+75.0%*

Tissue Analysis, consistent macro and micro mineral increases

Leaf portion	2 applications	3 applications	
Nitrogen	+10.5%	+19.3%	(indicates protein)
Magnesium	+2.3%	+13.6%	(indicates chlorophyll)
root portion			
Magnesium	+11.5%	+11.5%	(indicates chlorophyll)

Dry Matter *(roots)*	Control	2 applications	3 applications
% Dry Matter Content	12.46%	16.17%	14.18%
% increase over control		*+29.78%*	*+13.8%*

Harvest yield weights *from 10 linear feet of each category (does not include unmarketable beets)*

Weight (in lbs.)	Control	2 applications	3 applications
Large	9.95 lbs/10 pieces	17.41 lbs/16 pieces	16.81lbs/13 pieces
Medium	33.14 lbs/91 pieces	39.5 lbs/102 pieces	64.73 lbs/100 pieces
Small	16.58 lbs/103 pieces	39.11 lbs/167 pieces	31.91 lbs/173 pieces
Total Pieces	204	285	286
Total Weight	59.67 lbs	96.02 lbs	113.45 lbs
Marketable % increase (over control)		**+60.92%**	**+90.13%**

By mistake the field workers for this farm went through the all test plots once and harvested the largest beets. This reported harvest is the equivalent of a 2nd picking. The first picking would probably have given even greater yields for the treated samples.

County Line Farm Baby Carrot Harvest Results March 2012 (report summary)

Drench date - 10/20/2011 Seed date- 10/19/2011

Emergence- 10/26/2011

1st app date - 11/16/2011, 2nd app date - 12/7/2011, harvest date – 2/13/2012

Duration of growing cycle – 118 days

Soil pH 7.95

Brix Refractometer readings *(average of 4 plots)*

Control	1 application	2 applications
7.0	9.2	8.95
% increase over control	*+31.4%*	*+27.9%*

Tissue Analysis, consistent macro and micro mineral increases

% increase over control	1 application	2 applications
Calcium (leaf)	+ 3.2%	+ 10.2%
Calcium (root)	+5.9%	5.9%

Harvest yield weights *in pounds selected from 2 of the 4 beds.*

Weight (in lbs.)	Control	1 application	2 applications
Quality Culls	3.04	2.13	3.31
Marketable	72.41	90.78	102.51
Total	75.45	92.91	105.82
Marketable % increase (over control)		**25.37%**	**41.57%**

Harvest weights of immature carrots do not truly represent total harvest weights from the entire growing season if the carrots were allowed to mature.

Insect Damage

% decrease from the control	1 application	2 applications
Decrease	33%	33%

County Line Farm Cauliflower Harvest Results February 2012 (report summary)

Transplant date - 10/20/2011, Soil drench date - 10/21/2011
2nd app date - 11/17/2011, 3rd app date - 12/8/2011
harvest date - young harvest 1/31/2012 mature harvest (for LP test)-2/15/2012
Duration of growing cycle (seedlings transplanted at 6 weeks)- 145 days

Brix Refractometer readings *(flower)*

Control	2 applications	3 applications
9.0	9.5	10.0
% increase over control	*+5.5%*	*+11.1%*

Spectrographic analysis of roots *– Sea-Crop treated compared to control*

% increase over control	2 applications	3 applications
Calcium	+ 69%	+ 31%
Iron	+548%	+183%
Manganese	+224%	+46%
Silicon	+196%	+ 82%
Zinc	+284%	+ 9%

Tissue Analysis, consistent macro and micro mineral increases

% increase over control	2 applications	3 applications	
Nitrogen (leaf)	+ 16.4%	+ 7.0%	(indicates protein content)
Potassium (Petiole)	+1.25%	+.94%	
Calcium (petiole)	+15.0%	+ 6.5%	
Zinc (leaf)	+.4%	+2.29%	
Boron (leaf)	+6.3%	+2.3%	

Root Ball Dry Matter

% increase over control	2 applications	3 applications
weight	+ 7.6%	+10.0%

Flower Size *(diameter)*

% increase over control	2 applications	3 applications
diameter	+ 18.96%	+25.73%

Flower Dry Matter *(equals increased nutrient density)*

% increase over control	2 applications	3 applications
weight	+21.3%	+24.3%

Flower Weight *(equals increased harvest)*

% increase over control	2 applications	3 applications
weight	**+72.9%**	**+89.5%**

Insect Damage *(holes per plant)*

% decrease over control	2 applications	3 applications
decrease	+15.0%	+40.5%

County Line Farm Fennel Harvest Results March 2012 (report summary)

Drench date – 10/20/2011 Seed date- 10/19/2011
Emergence- 10/25/2011
1st app date - 11/15/2011
2nd app date – 12/6/2011
3rd app date- 12/27/2011
Harvest date – 3/20/2012
Duration of growing cycle – 147 days

Brix Refractometer readings

Control	2 applications	3 applications
7.0	9.0	11.0
% increase over control	*+28.6%*	*+57.1%*

Tissue Analysis *(leaf),* consistent macro and micro mineral increases

% increase over control	2 applications	3 applications	
Nitrogen	+23.9	+ 42.1	(indicates protein content)
Potassium	+51.1	+39.3	
Magnesium	+7.4	+33.3	(indicates chlorophyll)
Iron	+68.6	+48.6	

Dry Matter *(green tops)*	Control	2 applications	3 applications
% Dry Matter Content	9.27%	9.25%	10.44%
% increase over control		*-.2%*	*+12.6%*

Harvest yield weights *in pounds selected from 1 of the 4 beds.*

Weight (in lbs.)	Control	1 application	2 applications
Large	80.34 lbs/55 pieces	83.22 lbs/57 pieces	111.32lbs/76 pieces
Medium	57.92 lbs/60 pieces	63.32 lbs/71 pieces	72.58 lbs/71 pieces
Small	7.83 lbs/17 pieces	16.00 lbs/29 pieces	18.79 lbs/31 pieces
Total Pieces	132	157	178
Total Weight	146.09 lbs	162.54 lbs	202.69 lbs
Marketable % increase (over control)		**+11.26%**	**+38.74%**

County Line Farm Baby Spinach Harvest Results March 2012 (report summary)

Drench date – 1/24/2012 Seed date- 1/22/2012
Emergence- 1/26/2012
1st app date - 2/15/2012
2nd app date – 3/1/2012
3rd app date- 3/12/2012
harvest date – 3/19/2012
Duration of growing cycle – 57 days

Brix Refractometer readings *(average of 4 plots)*

Control	2 applications	3 applications
5.25	6.75	6.5
% increase over control	*+28.6%*	*+23.8%*

Tissue Analysis (leaf), consistent macro and micro mineral increases

% increase over control	2 applications %	3 applications %	
Nitrogen	*+ 8.1%*	*+11.8%*	(indicates protein content)
Potassium	*+ .6%*	*+5.6%*	
Magnesium Content	*+18.3%*	*+19.5%*	(indicates chlorophyll)
Copper Content	*+18.3%*	*+7.7%*	

Weight in Grams *of the greens portion of 35 representative plants of each category.*

	Control	2 applications	3 applications
Fresh Weight	221	316	376
% increase over control		***+ 43.0%***	***+70.1%***
Dry Matter Weight	30	48	54
% Dry Matter Content	13.6%	15.2%	14.4%
% increase over control		***+11.8%***	***+5.9%***

At the request of the laboratory doing the tissue analysis (Fruit Growers Laboratory Inc.) this test was allowed to pass its harvest date by one week. As a result much of the actual harvest was rejected as too mature to meet the grower's standard for baby spinach.

County Line Farm Swiss Chard Harvest Results March 2012 (report summary)

Direct seed date - 10/17/2011
Drench date - 10/18/2011
Emergence- 10/22/2011
2nd app date - 11/15/2011
3rd app date - 12/6/2011
Harvest date - mature harvest 2/05/2012
Duration of growing cycle - 112 days

Brix Refractometer readings *(average of 4 plots)*

Control	2 applications	3 applications
9.5	9	11
% increase over control	-4.7%	+15.8%

Tissue Analysis *(petiole)*, consistent macro and micro mineral increases

% increase over control	2 applications	3 applications	
Calcium	+17.7%	+26.3%	
Magnesium	+9.0%	+36.4%	(indicates chlorophyll)
Potassium	+3.7%	+ 5.6%	
Sodium	+36.9%	+48.1%	

Dry Matter *(leaf)*

% increase over control	2 applications	3 applications
weight	+ 7.7%	+110.3%

Harvest Yield Weights *(from a single cutting)*

Weight (in lbs.)	Control	2 applications	3 applications
Quality Culls	63.6	56.56	54.46
Firmness Culls	48.42	22.49	44.56
Marketable	27.6	46.18	40.14
Total	139.62	125.23	139.16
Marketable % Increase (over control)		**+ 84.6%**	**+45.5%**

Harvest weights from a single cutting do not truly represent total harvest weights from the entire growing season.

Insect Damage *(holes per plant)*

% decrease from the control	2 applications	3 applications
Decrease	-33%	- 33%

The following result from an experiment with mice performed in 2003 show, the kinds of dramatic effects that direct nutritional supplementation with seawater concentrates can have.

Animal Research on Mice

Conducted in 2003

The Stamina test:
Forced swim testing to the point of terminal exhaustion.

The Method:
Four week old white mice were obtained and split into two populations. The control population consisted of ten mice and the other group consisted of thirty mice.
Both populations were given measured amounts of the same food and measured amounts of water daily. The test group was given a dose of seawater mineral extract in their water ration.
The maximum recommended oral dose for humans is 0.2 ml per kilogram of body weight per day.
The mice were each given 0.5 ml of the formula in their water every day for thirty days.
At the termination of the test the mice weighed an average of 27 grams each. The dosage which they had received for 30 days was equal to 9.25 ml per kilogram of body weight per day or more than 45 times the recommended dose for humans.
The 30 day duration of the test equals 1/24th of the life span of a mouse or the equivalent of a 3 and 1/3 year segment of an eighty year human life span.
During the first week of the test there was 30% mortality for the control population but no deaths occurred in the group receiving supplementation.

Termination:
At the end of 30 days the mice were placed in 1,000 milliliter beakers containing 800 milliliters of water where they were forced to swim until the point of terminal exhaustion and death.

The Results:
1. The control group, which had not received the mineral supplementation, endured for an average of 257 minutes, or 4 hours and 17 minutes.
2. The test population, which had been supplemented with the seawater mineral extract endured for an average of 840 minutes, or 14 hours.

Observation:
The test population, that received supplementation, exhibited more stamina than the control group by enduring the forced swim testing for a duration 3.27 times greater.

Conclusions:
1. The seawater extract used for this study greatly enhanced stamina in the test subjects.
2. Dietary supplementation with seawater mineral extract, at the dosage used, appeared to have only beneficial effects.

In regards to this study I would like to make a clarification. Mice float. Oils on their fur enable the fur to trap air which gives buoyancy. As the mice spend time in water the oils eventually are washed away and the natural buoyancy becomes reduced to the extent that the mice then must start to exert themselves by swimming in order to stay afloat. The treated population in this experiment floated for a much longer period of time than the controls before having to exert themselves. The indication is that the treated mice had much healthier coats of fur and that made a major contribution to the results noted above.

This work took place just before I started doing agricultural experimentation. At this point in time my focus had been making seawater concentrate as a mineral supplement for human consumption

Sacrificing the mice in terminating this experiment was an unpleasant task so I did no further work along these lines.

APPENDIX B

Composition of Seawater

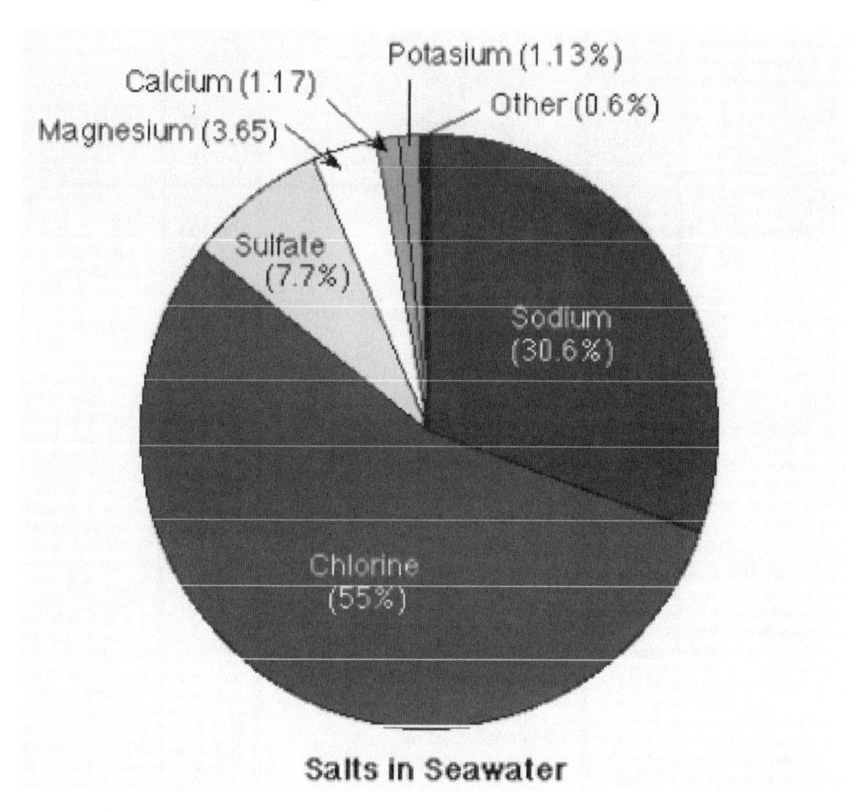

Salts in Seawater

Detailed composition of seawater at 3.5% salinity

Element	Atomic weight	ppm	Element	Atomic weight	ppm
Hydrogen H	1.0079	110,000	Molybdenum Mo	0.09594	0.01
Oxygen O2	15.999	883,000	Ruthenium Ru	101.07	0.0000007
Sodium Na	22.989	10,800	Rhodium Rh	102.905	.
Chlorine Cl	35.453	19,400	Palladium Pd	106.4	.
Magnesium Mg	24.312	1,290	Argentum (silver) Ag	107.870	0.00028
Sulfur S	32.064	904	Cadmium Cd	112.4	0.00011
Potassium K	39.102	392	Indium In	114.82	.
Calcium Ca	10.080	411 67.3	Stannum (tin) Sn	118.69	0.00081
Bromine Br	79.909		Antimony Sb	121.75	0.00033
Helium He	4.0026	0.0000072	Tellurium Te	127.6	.
Lithium Li	6.94	0.170	Iodine I	166.904	0.064
Beryllium Be	9.0133	0.0000006	Xenon Xe	131.30	0.000047
Boron B	10.811	4.450	Cesium Cs	132.905	0.0003
Carbon C	12.011	28.0	Barium Ba	137.34	0.021
Nitrogen ion	14.007	15.5	Lanthanum La	138.91	0.0000029
Fluorine F	18.998	13	Cerium Ce	140.12	0.0000012
Neon Ne	20.183	0.00012	Praesodymium Pr	140.907	0.00000064
Aluminium Al	26.982	0.001	Neodymium Nd	144.24	0.0000028
Silicon Si	28.086	2.9	Samarium Sm	150.35	0.00000045
Phosphorus P	30.974	0.088	Europium Eu	151.96	0.0000013
Argon Ar	39.948	0.450	Gadolinium Gd	157.25	0.0000007
Scandium Sc	44.956	<0.000004	Terbium Tb	158.924	0.00000014
Titanium Ti	47.900	0.001	Dysprosium Dy	162.50	0.00000091
Vanadium V	50.942	0.0019	Holmium Ho	164.930	0.00000022
Chromium Cr	51.996	0.0002	Erbium Er	167.26	0.00000087
Manganese Mn	54.938	0.0004	Thulium Tm	168.934	0.00000017
Ferrum (Iron)	55.847	0.0034	Ytterbium Yb	173.04	0.00000082
Fe	58.933	0.00039	Lutetium Lu	174.97	0.00000015
Cobalt Co	58.710	0.0066	Hafnium Hf	178.49	<0.000008
Nickel Ni					
Copper Cu	63.54	0.0009	Tantalum Ta	180.948	<0.0000025
Zinc Zn	65.37	0.005	Tungsten W	183.85	<0.000001
Gallium Ga	69.72	0.00003	Rhenium Re	186.2	0.0000084
Germanium Ge	72.59	0.00006	Osmium Os	190.2	.
Arsenic As	74.922	0.0026	Iridium Ir	192.2	.
Selenium Se	78.96	0.0009	Platinum Pt	195.09	.
Krypton Kr	83.80	0.00021	Aurum (gold) Au	196.967	0.000011
Rubidium Rb	85.47	0.120	Mercury Hg	200.59	0.00015
Strontium Sr	87.62	8.1	Thallium Tl	204.37	.
Yttrium Y	88.905	0.000013	Lead Pb	207.19	0.00003
Zirconium Zr	91.22	0.000026	Bismuth Bi	208.980	0.00002
Niobium Nb	92.906	0.000015	Thorium Th	232.04	0.0000004
			Uranium U	238.03	0.0033
			Plutonimu Pu	(244)	.

Note! ppm= parts per million = mg/litre = 0.001g/kg.
source: Karl K Turekian: *Oceans*. 1968. Prentice-Hall

APPENDIX C

ORAC Assays

In these ORAC studies, SEA-CROP® is referred to as Immutines. In Northern Europe, Hak Agro Feed of The Netherlands markets SEA-CROP® under the trade name "Immutines."

Appreciation is given to Hak Agro Feed for all of the fine research that they have done to help develop an understanding of how best to utilize seawater concentrates for the benefit of both plant and animals.

TEST RESULTS (1):

Samples	ORAC value (1Jmol TE /100 g.)*	
Control cucumbers (C)	90.0 ± 1.) (mean ±S.D.)	(n = 4)
Immutines-treated cucumbers (3 ml/m²/week, B)	123.3 ± 15.8 (mean± S.D.)	(n = 4)
Immutines-treated cucumbers (4 ml/m²/week, R)	116.7 ± 8.7 (mean ±S.D.)	(n = 4)

**As stated previously, ORAC values are expressed as pmol TEper 100 g of test sample. If required, customer can extrapolate these ORAC values to pmol TEper cucumber or pmol TE per serving.*

NB: 1 Jlmol Trolox Equivalents (TE) equals 250 JlY Trolox

	Antioxidant capacity (%of	Increase in antioxidant capacity (relative to control)
Control cucumbers (C)	100	-
Immutines-treated cucumbers (3 ml/m²/week, B)	137 ± 17.6	37.0 %
Immutines-treated cucumbers (4 ml/m²/week, R)	129.7 ± 9.6	29.7 %

	Antioxidant capacity (% of control)	Increase in antioxidant capacity (relative to control)
Control cucumbers (C)	**100**	-
Immutines-treated **cucumbers (1 ml/m², G)**	**136.63 ± 21.5**	**36.6 %**
Immutines-treated **cucumbers (1.5 ml/m², R)**	**133.3 ± 36.1**	**33.3 %**

Responsible Investigator:

Dr. E. van den Worm
(CEO, ORAC Europe BV)

(e-mail: E.vandenworm@uu.nl)

• ORAC Europe BV • PO BOX 80082 • 3508 TB Utrecht • The Netherlands •
• Rabobank: 12 77 66 456, Utrecht • IBAN: NL41RABO 0127 7664 56 •
• SIC: RABONL2u • KvK 30222065 • VAT nr. NL8174.77.615.801 •
• Tel: 00 31(0)302535933 • Fax:00 31(0) 302536941 •
• e-mail: info@orac-europe.com •
• www.orac-europe.com
•

	Antioxidant capacity	Increase in antioxidant capacity (relative to control)
Control tomatoes (C)	**1**	-
***Immutines-treated* tomatoes (1 mllm²/week, I)**	**113.6 ± 2.5**	**13.6 %**

Responsible Investigator: Dr. E. van den Worm (*CEO,* ORAC Europe BV)

• ORAC Europe BV • PO BOX 80082 • 3508 TB Utrecht • The Netherlands •
• Rabobank: 12 77 66 456, Utrecht • IBAN: NL41RABO 0127 7664 56 •
• BIC: RABONL2u • KvK 30222065 • VAT nr. NL8174.77.615.801 •
• Tel: 00 31(0)302535933 • Fax:00 31(0) 302536941 •
• email: info@oraceurope.com •
• www.orac-europe.com •

Index

NOTES

1: What is Seawater?

1. "NASA Satellite Sees Ocean Plants Increase, Coasts Greening" *Science Daily,* (Mar. 9, 2005)

2. Jed A. Fuhrman, "Marine viruses and their biogeochemical and ecological effects", *Nature 399, 541-548,* (10 June 1999)

3. Herbert Swenson, "Why Is the Ocean Salty?" *Geological Survey (Dept. of Interior),* (1983)

4. Katina Bucher Norris, "Dimethylsulfide Emission: Climate Control by Marine Algae?" *Aquatic Sciences and Fisheries Abstracts,* (Nov., 2003)

2: A History of Seawater Concentrate Research

1. Maynard Murray, M.D., "Sea Energy Agriculture" *Acres U.S.A.,* (2003)

2. Maynard Murray, "Process of Applying Sea Solids as Fertilizer" US Patent # 3,071,457, *U.S. Patent and Trademark Office,* (Jan., 1, 1963)

3: The Benefits of Seawater Concentrate

1. Maynard Murray, M.D., "Sea Energy Agriculture" *Acres U.S.A.,* (2003)

2. Jonathan D. Kaplan, "Managing Manure in California's Central Valley' *Dept. of Economics, California State University Sacramento,* PDF file referenced May 2, 2012

4: Nutrient Density

1. William A. Albrecht, Ph. D., "Soil Fertility & Animal Health", *Acres U.S.A.,* (2005)

2. Ibid.

3. Maynard Murray, "Process of Applying Sea Solids as Fertilizer" US Patent # 3,071,457, *U.S. Patent and Trademark Office*, (Jan., 1, 1963)

5: How Plants Grow

1. "Metalloprotein" *Wikipedia*, referenced (May 2, 2012)

6: Hydroponics

1. Maynard Murray, M.D., "Sea Energy Agriculture" *Acres U.S.A.,* (2003)

9: How Plants Shouldn't Grow.

1. Maureen O'Hagan and Kristi Heim, "Gates Foundation ties with Monsanto under fire from activists" *Seattle Times,* (August 28, 2010)

2. Chad Heeter, "Seeds of Suicide" *Frontline Video Report*, (July 26, 2005)

Alex Rossi, " Wave Of Suicides Among Indian Farmers" *U.K Sky News* ,(November 18, 2011)

Andrew Malone, "The GM genocide: Thousands of Indian farmers are committing suicide after using genetically modified crops" *U.K. Daily Mail*, (2 November 2008)

3. Dr. Don Huber, "What's New in Ag Chemical and Crop Nutrient Interactions" *Fluid Journal Vol. 18 No. 3, Issue #69*, (Spring 2010)

4. Caroline Cox, "Glyphosate Factsheet" / *Journal of Pesticide Reform v.108, n.3*, *(*Fall98 rev.Oct00)

5. Dr. Don Huber, "What's New in Ag Chemical and Crop Nutrient Interactions" *Fluid Journal Vol. 18 No. 3, Issue #69*, (Spring 2010)

6. Ibid.

7. Caroline Cox, "Glyphosate Factsheet" / *Journal of Pesticide Reform v.108, n.3*, *(*Fall98 rev.Oct00)

8. " Monsanto guilty in 'false ad' row" *B.B.C. News*, (15 October 2009)

"Glyphosate" *Wikipedia,* referenced (May 2, 2012)

9. Ibid.

"Roundup (herbicide)" *Wikipedia,* referenced May 2, 2012

10. Jeffrey M. Smith, "Monsanto's Roundup Triggers Over 40 Plant Diseases and Endangers Human and Animal Health" *Natural News,* January 28, 2011

11. Ibid.

12 Dr. Don Huber, "What's New in Ag Chemical and Crop Nutrient Interactions" *Fluid Journal Vol. 18 No. 3, Issue #69*, Spring 2010

13 "Widely Used Herbicide Commonly Found in Rain and Streams in the Mississippi River Basin", *Technical Announcement, USGS Newsroom,* 8/29/2011

14. Caroline Cox, "Glyphosate Factsheet" / *Journal of Pesticide Reform v.108, n.3*, Fall98

Lucia Graves "Roundup: Birth Defects Caused By World's Top-Selling Weedkiller, Scientists Say" *Huffington Post*, 08/24/11

Resources

Sea Energy Agriculture by *Maynard Murray, M.D.* published by *Acres U.S.A.*, P.O. Box 91299, Austin, Texas 78709 (512) 892-4400, 1-800-355-5313. www.acresusa.com

These articles concerning glyphosate toxicity may all be found on the internet:

Fluid Journal Fall 2007 Vol. 15, No. 4 - Issue # 58 - Pages 20-22
What About Glyphosate-Induced Manganese Deficiency? ~ *Dr. Don M. Huber*
http://www.fluidfertilizer.com/pastart/pdf/58P20-22.pdf

Fluid Journal Spring 2010 Vol. 18 No. 3, Issue #69 ~ *Dr. Don Huber*
http://www.fluidjournal.org/1gsdgfs-S10/S10-A4.pdf

– Current Update ~ *Don M. Huber*
http://www.fluidjournal.org/1gsdgfs-S10/S10-A4.pdf

Glyphosate Induced Hidden Hunger ~ *Davidson, D.* 2010
http://fhrfarms1.com/education.php

Glyphosate endocrine disruption& genetic damage:

http://en.wikipedia.org/wiki/Roundup_(herbicide)

http://www.ncbi.nlm.nih.gov/pubmed/19539684

http://www.gmwatch.eu/latest-listing/1-news-items/13631-now-glyphosate-found-in-peoples-urine

Roundup and Birth Defects:

Reported in the Huffington Post June 24, 2011 ~ *Lucia Graves*
http://www.huffingtonpost.com/2011/06/24/roundup-scientists-birth-defects_n_883578.html

Indian Farmer Suicide:

Reported in the U.K. Daily Mail, Nov. 2, 2008 ~ *Andrew Malone*
http://www.dailymail.co.uk/news/article-1082559/The-GM-genocide-Thousands-Indian-farmers-committing-suicide-using-genetically-modified-crops.html

Wave Of Suicides Among Indian Farmers, Reported in the U.K Sky News November 18, 2011 ~ *Alex Rossi*
http://news.sky.com/home/world-news/article/16112805

Seeds of Suicide-Frontline Video Report ~ *Chad Heeter* July 26, 2005 by
http://www.pbs.org/frontlineworld/rough/2005/07/seeds_of_suicid.html

Monsanto false advertising and scientific fraud:
http://en.wikipedia.org/wiki/Roundup_(herbicide)

Acknowledgements:

Thanks are due to the following individuals and companies who at their own considerable expense sponsored formal third party testing of SEA-CROP®.

Bruce Kemper of Cal-Agri Products, LLC

Barend Hak of Hak Agro Feed

Jeffery Collé of Collé Agriculture, LLC

Thanks are given to Fred Walters of Acres USA for his kind permission to reproduce Dr. Maynard Murrays test data from Sea Energy Agriculture by *Maynard Murray, M.D.* originally published by *Acres U.S.A.,* P.O. Box 91299, Austin, Texas 78709 (512) 892-4400, 1-800-355-5313.

Thanks are also given to the distributors and customers of Ambrosia Technology products for their commercial support.

Thanks are given to Jeff Gee whose thorough and creative critical editing of the manuscript has made this a much better book than it would have otherwise been. However, the author reserves for himself the credit for any errors that it may yet contain.

Last but not least thanks are given to the author's family members, especially his wife, who both supported and endured years of research and development.

About the Author

Arthur Zeigler has been a precious metal assayer, precious metal refiner, prospector, mine operator, mining consultant, laboratory director and both a metallurgical and agricultural researcher.
He has spent nearly four decades doing independent research in extractive metallurgy.
Current research is devoted to the use of trace minerals for improvement of agricultural soils and enhancement of both plant and animal health.
He lives in the Pacific Northwest.